Adaptive Cooperation between Driver and Assistant System

Frédéric Holzmann

Adaptive Cooperation between Driver and Assistant System

Improving Road Safety

With 135 Figures and 2 Tables

Springer

Frédéric Holzmann
Siemens VDO Automotive AG
Group Strategy / Technology
Siemensstrasse 12
93055 Regensburg
Germany
holzmann.frederic@siemens.com
www.siemensvdo.com

ISBN: 978-3-642-09388-3 e-ISBN: 978-3-540-74474-0

Cover design: Erich Kirchner, Heidelberg

Printed on acid-free paper

9 8 7 6 5 4 3 2 1

springer.com

To all my family
and the SPARC partners
for their unvaluable support

Preface

As I started writing this book, I was remembering two particular moments in my life. The first one happened a few years ago as I fell asleep in my car while driving by night. Me and my friends thought that driving after a heavy mountaineering day was not dangerous (even when drowsy). As we escaped the accident miraculously without any casualties, I knew that I would be working on all my life in the name of all the other people, who would not be so lucky as we were.

A few years later I was participating at the kick-off meeting of the research project named SPARC, which was partially funded by the European Commission as part of the sixth framework program. At this event we acknowledged that we would have to work on a new fully integrated vehicle architecture. No one knew how far these investigations would be performed. Today, I can appreciate the results of these 3 years.

All the new concepts, which have been integrated into those prototypes, are born out of several iterations which hardly converged. Anyhow, this document sums up the logical concept that has been worked on step by step leading to those final results.

Now our next step will be to improve the substitution capacity for the driving task on top of this overall vehicle architecture. Hence we will show the next step by having a highly automated and reliable safe vehicle platform with still the dream of full autonomous vehicles in the background.

Regensburg, September 2007 *Frédéric Holzmann*

Contents

Part IV Adaptive cooperation

Part I

New concept of cooperation

1

Needs of improved assistant systems

1.1 Analysis of the cause of accidents on the road

Everywhere around the world, safety primarily depends on human factors. Literature survey has pointed out driver impairment as the dominating cause of all traffic accidents. This ratio has been estimated above 90% in countries such as the USA [93], Europe [115] or Japan [123] which have new vehicle technologies. For countries like eastern Europe, this percent goes down to 60% even if the decrease of road victims has reached 50% per year, or lower, for China, India or Africa, where the percent of older vehicles is quite high, meaning a higher risk of mechanical defect.

Blincoe and Faigin presented in a survey [13] the cost of the motor-vehicle crashes during 1990 in the USA. They accounted for 44,531 fatalities, 5.4 million non-fatal injuries, 28 million damaged vehicles and an estimated total cost of 137.5 billion $. Most costs were due to personal property damage (45.7 billion $ (33%)), productivity losses in the workplace (39.8 billion $ (29%)), medical-care expenses (13.9 billion $ (10%)) and losses related to household productivity (10.8 billion $ (8%)). Thus, the economic impact of motor-vehicle crashes during 1990 was approximately 2.5% of the gross domestic product in the USA.

Moreover, data show that highway incidents, including collisions, rollovers and other traffic mishaps, remained the leading cause of work-related deaths in 2002, accounting for 25% of the occupational fatalities in the USA. Additionally, truck driving remains one of the most dangerous jobs in America, with 69% of those driver fatalities resulting from highway incidents [96].

Furthermore an insufficient maintenance or inadequate tire pressure leads to vehicle failure and they provoke accidents with the same percents. The first cause will be downsized over the years with the introduction of new vehicles fitted with new technology. And for the second cause, the vehicle manufacturers are integrating pressure sensors into the tires to prevent the risk of explosion. However, the real challenge remains in dealing with the other 95%, which are strictly related to the driver's behavior.

After dedicated accident investigations such as inattention in [122], or fatigue in [57] for trucks, lateral collisions in [85], trucks on freeways in [48], collisions implicating trucks and non-trucks actors in [25], etc., loss of alertness or indirectly recognition errors (perception, comprehension, delays and decision errors) has been determined as predominant factor of fatal accidents (34%). The impact of the lack of concentration will be lowered by 30% if the driver reacts 500 ms earlier as shown by Lunenfeld [79] and Egeth [36]. Therefore the role of the assistant systems are to avoid accidents by giving early advices to the driver in order to save this 500 ms or to correct temporarily the vehicle command to lower the impact of these dangerous situations.

Parallel to the effect of the driver drowsiness, statistics shows interesting similarities between the different accidents. Simple data mining such as in figure 1.1 extracts several standard accident scenarios for personal cars and trucks, in urban environment or outside cities. Four main sources of accidents, which are related to more than 60% of the overall accidents, can be defined and used for the first requirements of assistant systems. However, getting a 100% safe vehicle is still a dream as it was in aeronautics for 20 years. This goal cannot be reached as sometimes accidents are fully unforseenable and cannot be avoided:

1. *Collision with the vehicle ahead on highway*: with standing or moving road-users. In most instances, it is due to wrong analysis of the distances, speed difference and road adherence.
2. *Collision with a crossing vehicle in the city or speedway*: most of the time it is due to a problem of synchronization or driving rules misunderstood/misused. Vehicles cannot stop anymore while another is turning.

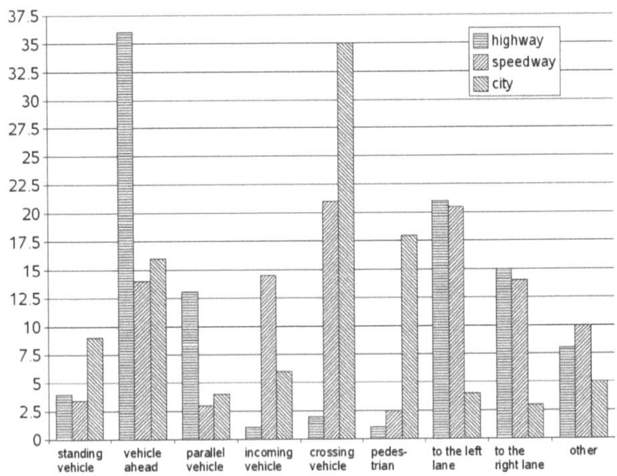

Fig. 1.1. Percent of accidents on the road for the year 2004 in Europe

3. *Collision while lane changing on highway*: problem with dead angle in most cases. Such accidents hit personal cars most of the time; for trucks other vehicles take more care of distances and speeds because of the fear on seeing truck lane merging.
4. *Accident with a pedestrian*: 70% of pedestrian are in front of the vehicle during the collision and more than 80% are moving.

First of all, the fundamental differences between all these scenarios request adapted sensors strategies for the surrounding environment. Moreover different approaches are required to solve each problem. For the first and third scenarios, dedicated devices have already been integrated into vehicles as the scenarios are well defined and the actions of the assistant systems are not too complex. For the two other scenarios, the complexity of the situation and the need for a good sensing and inferring has delayed the introduction of robust technology on the market.

1.2 Autonomous vehicles as possible solution

All around the world, the revolution in the computation capacity has given the possibility for research projects to tackle autonomous vehicles. IVHS projects such as PATH [114] in California or in Europe with Prometheus [31], Ralph [103], Gold [10] try to skip the human intervention by developing full autonomous vehicle, from the environment perception down to the actuators control. In spite of promising results from the different institutes on open highways, these solutions have still limited cognition capacities. Cluttered urban roads were proven difficult to cope with autonomous systems due to the increase in road scene complexity [12]. On urban situations, human drivers rely much more on behavior prediction of the other road users and pedestrians. This high-level probabilistic reasoning cannot be easily modeled and solutions based on artificial intelligence still cannot be integrated onto embedded platforms [8].

Up to now, the algorithms dealing with autonomous vehicles share the same patterns. First a split between static environment and dynamic environment is done. This distinction comes purely from the history as such systems were only dealing with static environment at the beginning. Hence the dynamic tracking of the objects can be seen as an extension of the algorithms. Second, one of the primary tasks (global and local path plannings) is used at a certain time depending on the dangerousness (or the time to collision TTC) or the computation capacity available (figure 1.2).

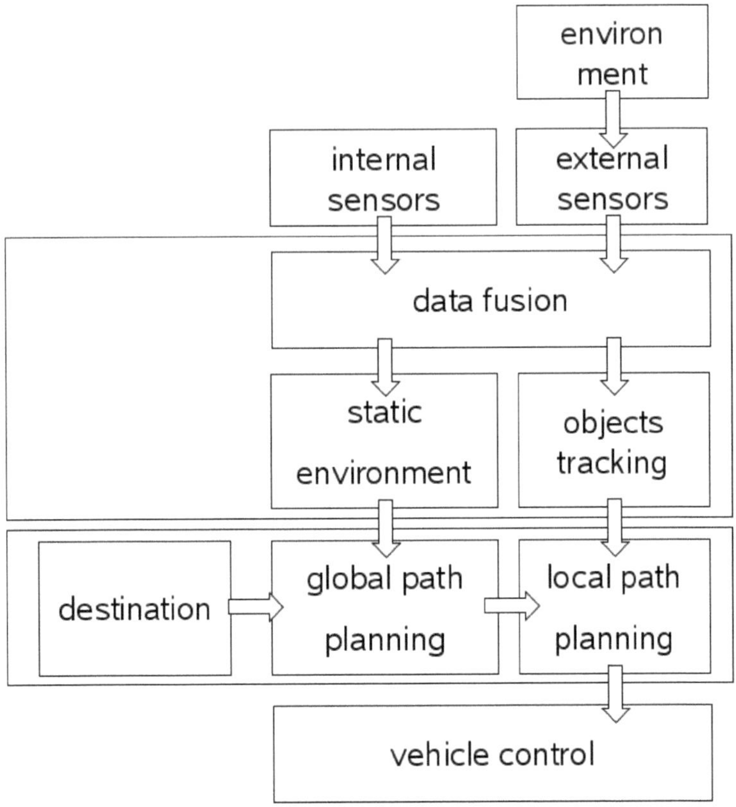

Fig. 1.2. Two-layer architecture model for autonomous vehicles

1.3 Ways for improving the driving safety

In 2001, the European Commission for Road Safety has taken the initiative to publish the European Road Safety Charter [41]. This charter has the ambitious objective to decrease the number of fatalities on the European road by half till 2010.[1] As for other important projects around the world, the project works on three different ways of improvement (presented here after: infrastructure, driver capacities, intelligent vehicles) by supporting research and development through integrated projects led by industries. The trends in such projects have always been to try to integrate all the possible technologies with one dedicated point of view. Thus no real potential full architecture for infrastructures or vehicles has been defined as a norm. Such standardization will be possible only if a clear re-organization of the functionality modules is performed.

[1] At this date, the eastern part of the Europe was not integrated into the Union. This extension makes the achievement of the goal statistically impossible.

1.3.1 Improvement of the infrastructures

Some roads are particularly dangerous. This depends most of the time on the weather (risk of aquaplaning, etc.) and on the vehicle density (roads junctions, etc.). After a methodological analysis of the annual road accident reports of 2002 [96], these sites have been isolated and new solutions have been simulated. As as result, road re-organization has been planned to lower this inherent risk. On top of this preventive work, additional help can be provided in reaction of events. Thus data on the road can be gathered continuously (vehicle density, snow, fog, etc.) and relevant information can be provided to the drivers by use of intelligent traffic signs or info-cars.

1.3.2 Improvement of the driver capacities

Even if the theoretical risk on the road is lowered with the improvements presented previously, safety will still need to be improved through driver monitoring and driver conditioning. On one side, driver fitness sensing continuously evaluates directly or indirectly the driver behavior and will inform the driver on his/her alertness as soon as it starts sinking. In Europe, most researches on this topic have been realized during the AWAKE project [102]. On the other side, European projects such as Aide [2] are dealing with the problems set of providing correctly the information to the driver to help him/her to correct actively the vehicle command. But in most cases, even if the driver knows that he/she is drowsy, he/she will drive anyway because of mandatory obligations like reaching home or an appointment. Thus such functionality will only help the driver to know his/her level of activity but it is not enough to improve the safety.

1.3.3 Improvement of the vehicles

For 15 years, the car manufacturers have started to integrate first ABS and later ESC (stability control) systems into vehicles. These functionalities improve the realization of the command because they can avoid inadequate vehicle command in more than 74% of the cases by correcting temporarily this command. But when the functions have to go one step further, the complexity of their perception and cognition task will exponentially increase. In Europe several projects have been undertaken such as Prometheus, and later the *invent* initiative in Germany or the European Prevent project. But at each time, the assistant systems are only considered as comfort functions which have to be used under driver control, who is always taking the full responsibility of their actions.

1.4 Introduction of assistant systems and inherent problems

1.4.1 Current integration of assistant systems

The integration of driver assistant systems is performed step by step with budget constraints. Thus, the integration has been done as close as possible to the actuators to save bandwidth on the internal buses. This is the reason why the two first active systems already integrated into the vehicles are adapting the distance to the vehicle ahead (ACC), using one long-range radar, and is directly integrated into the engine management, and the semi-automatic parking assistant that controls the steering wheel. Here one vertical layer integration is defined (like in figure 1.3) when all the actuators are already connected to one assistant system. Thus the new assistant systems are either purely passive (just connected to the HMI) or an intensive communication between all assistant systems is required for their synchronization. The second possibility has also drawbacks since it is only validated for one typical vehicle topology and it has to be validated again for each new configuration. Moreover the synchronization of the longitudinal and lateral controls is not easy without intensive communication between all controllers. Hence, some active assistants such as the lane changing are difficult to integrate without huge bandwidth. Over the time, activities have been shown to solve this problem. The most common solution leads to a software integration of the different driver assistant systems since it is easier to have communication between software modules as between physical components over a physical network.

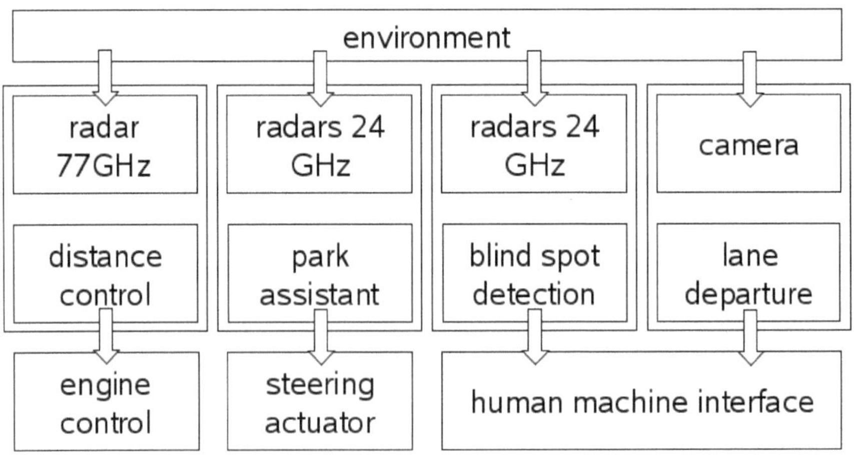

Fig. 1.3. Vertical layering

1.4.2 New issues coming from assistant systems

The increase of the complexity in the cockpit has a direct effect on the probability of failure by the components [64] and errors by the driver. The amount of devices installed in the vehicle needs additional alertness of the driver, who has to monitor continuously their action, while his/her alertness is decreasing due to the automation of some tasks. Resulting from this conflict, reported first by Brookhuis and De Waard in [19], the need of a well-designed driver interface becomes more important. Inadequately designed devices can adversely affect the driving tasks or even reduce its acceptance [8]. On the other hand, assistant systems, which only supply information to the driver, are most likely to meet acceptance from the driver [131] because of their passive task.

1.4.3 Behavioral changes with the human supervision

The main objective of the assistant systems is to lower the driver workload by realizing some tasks safely. Humans still have to monitor these systems. Unfortunately humans are poor process monitors as reported by Molloy in [89]. First of all, reaction time increases as soon as different processes have to be monitored at the same time. Secondly, such continuous monitoring process increases the driver workload over the time [54].

1.4.4 Risks of complacency

Experiences in aeronautics such as from Baindridge [6] or Sheridan [112] have shown how crucial the human contribution is in difficult operations. Indeed experience of pilots will be more and more limited during the next decades because of automation. Thus, skills that have been formerly acquired from manual flight will be missing during flight back to manual control under pressure. This attitude of reliance on automated systems has been called *complacency* [134]. Research on adaptive task allocation shows that temporarily returning control to human operator has positive effect with respect to this attitude [99]. This phenomenon has also been detected by drivers on simulator. For example, Ward [133] found evidences for complacency by drivers having a vehicle equipped with distance control and lane departure warning.

However, this methodology of temporal control changing is not acceptable for non-professional drivers in vehicles. Indeed drivers of passenger cars do not want to get time to time the command back as long as it is not vital. Here another solution has to be found, where the driver has to be monitored indirectly.

1.5 Problem statement and improvements with SPARC

Up to now the assistant systems were only substitution devices for the driver or hints providers. However, the next big challenge will be to shift the vehicle behavior responsibility from the driver side to the vehicle control side. This radical change will imply an assistant system that has to be reliable in each situation and it will also imply a cooperation process to intervene optimally on the driver's command.

This document will present several improvements that have been integrated within the European SPARC project. First a new vehicle architecture will be presented in order to manage the new level of integration complexity, which is the source of potential failure. Then a new generic assistant systems framework will be presented to insure safety relevance and a first concept of cooperation process will be also investigated. The elements presented hereafter are new and disruptive with respect to the current concepts as the integration of the systems will lead to a new concept that is the vehicle safety envelope. Finally the driver interface will be drastically re-worked to manage this new information keeping the driver in the control loop through scenario-dependent and driver-dependent feedback.

The SPARC project was started in January 2004 for 3 years. Managed by DaimlerChrysler for the European Commission with support from Irion Management, this FP6 STREP has focused on a failure-tolerant and scalable driver-by-wire architecture, which can be applied from heavy good vehicles down to the smallest passenger cars. Then a virtual driver and a HMI have been plugged with the ability of self-adaptation on the vehicle type.

Trucks from DaimlerChrysler have been prepared with KnorrBremse and Haldex for brake-by-wire with stability control, PLC to the trailer from YAMAR, differential lock from MagnaSteyr ECS and tires from Michelin. The trailer has been prepared by Kögel Fahrzeugwerke and Georg Fischer. The passenger cars from DaimlerChrysler integrate brake-by-wire technology from SiemensVDO Automotive, a steer-by-wire from SKF.

All vehicles integrated an energy management from ESG and IQ-Power. The ECUs developed by Continental Teves integrate a Freescale PowerPC and the redundancy management from the University of Stuttgart (ILS).

A camera from the EPFL and navigation platform with eHorizon from SiemensVDO Automotive have been used for the co-pilot from SiemensVDO Automotive. Finally the HMI management has been developed conjointly with the German Aerospace Center (DLR) and SiemensVDO Automotive.

Prior tests have been performed at different levels. The ECUs and the operating system have been validated on ETAS test platforms. The vehicles have been tested on Dürr Assembly Products test benches. And the functionalities have been tested with drivers in simulators coming from Simtec, the University of Würzburg and from the Fiat Research Center.

2

Adaptive cooperation between driver and assistant system

Abstract. The introduction of assistant systems has two main drawbacks. First the technological limit obviates the realization of full autonomous vehicles and makes mandatory the actions dangerous from the driver in complex situations such as in urban places. Second the driver shifts more and more his/her task to the monitoring and stays just as a back-up for the technological solution. Unfortunately he/she does not master adequate request as the human leaves quickly out of the control loop. Moreover, the lack of practice may influence negatively their actions with respect to complacency risks.

Therefore it is necessary to change the interaction design of the couple driver–vehicle. Hence the machine-centered design has to be replaced by an integrated human–machine approach, which may give better global results. Instead of automating as much as possible, the strategy will be to automate just the necessary tasks at the right moments. The assistant systems will be combined to define a virtual driver. Its cooperation with the driver will be adapted in real time depending on both situation awareness, the driver's needs of cooperation and his/her activity level.

First of all the vehicle architecture based on a vertical layering (direct link of a dedicated sensor to an ECU for one single actuator) has to be changed radically. Therefore, the first action will be to re-organize this architecture by using a horizontal layering where all the functionalities are integrated within the main domains. Hence it will be possible to add new functionalities within the different layers without having to modify the overall architecture.

2.1 Vehicle architecture matching the driver cognition flow

As today, a dedicated electronic control unit is allocated per functionality in a vehicle. This approach, corresponding to the actual suppliers' business model, may not be the best solution. Indeed the addition of the new functionalities will make the integration of the devices more complex and less reliable (compromises between cost and efficiency). That is why an extensive re-design of the architecture should be done by splitting the hardware and the software. The system architecture will handle the shifting from hardware connections to software integrations.[1] For a clear integration, the software architecture can be inspired from the driver cognition, since it is the most compliant architecture for cooperation with the functions within the driver.

The set of primary tasks defined first by Rasmussen in [106] and improved through works of Suetomi [121], and Stanton [118], splits the intellectual activities of a driver in three main tasks levels (Fig. 2.1). The first tasks are driving related and, thus, safety relevant. It goes from vehicle stabilization up to navigation as described hereafter. The second task level is related to the improvement of the driving tasks, e.g., a driver needs to switch on the upper beam when he/she is driving into a tunnel. These tasks are not mandatory but they improve the driving reliability. Differing from the first two levels, the last tasks are not driving related at all, they are related to all the disturbances. They can be technology disturbances such as changing the radio channel or non-technology disturbances such as discussions with passengers. All in all, the driver has a maximal activity level capacity, which cannot cover all the tasks. Therefore, the driver focuses automatically on some tasks, those which are the most relevant in his/her opinion. It is possible to automate partially the second level tasks such as automatic beams or to manage technology disturbances such as radio when dangerous situation arises. Hence it is possible to get the driver more concentrated more on the most relevant tasks: the first level tasks.

The first level task as depicted in Fig. 2.2 integrates the driving related tasks. First a strategy function defines the road sections to reach a destination. Then a tactic function determines an adequate maneuver relative to the situation. Finally this action is transformed in terms of acceleration and

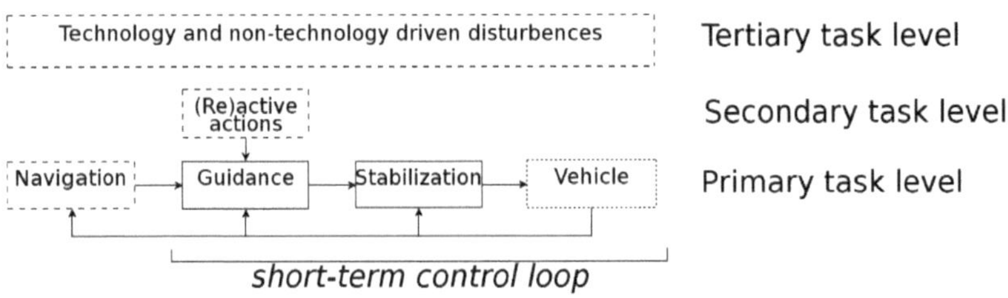

Fig. 2.1. Improved cognitive model for the driving tasks

[1] Moreover the costs of such architecture will be cheaper and more reliable because of the savings of wiring, connectors etc.

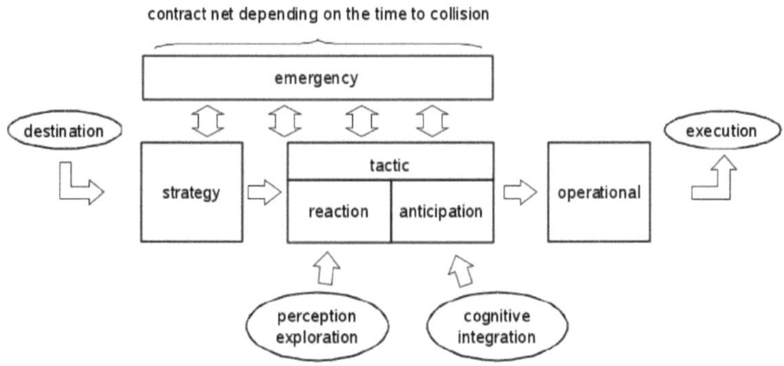

Fig. 2.2. Improved cognitive model for the driving tasks

steering angle, which will be transferred as a physical command to the vehicle through an operational function. A right integration of the assistant systems as a virtual driver has to be done in parallel to ensure a safe vehicle command even if the driver fails.

Two major points need to be improved to create an optimal human integrating architecture. First an emergency function is working in parallel to the strategy and tactic functions. This module will take the initiative to gain a hold on the other modules as soon as the driver perceives a danger. In this case the driver downsizes unconsciously his/her perception range and focuses only on the source of danger. That is why it is necessary to postpone the non-guidance functions while dangerous situation occurs.

Secondly the tactic function has to be improved through a split into two modules. The driver first has a look at all the possible maneuvers that he/she can perform as a reaction to the surrounding environment, and after that he/she will choose one of the maneuvers. Thus an advanced assistant system needs to deal with multiple maneuvers to get the capacity to react to the different needs of the driver. Moreover the driver can also face a dilemma as he/she has to anticipate situations such as an incoming curve. The driver will have to consider if overtaking is recommended, as the incoming curve requires speed degradation.

Two additional elements will be integrated into this cognitive flow even if their influence is low. On one hand humans have the capacity to remember locations and situations encountered earlier. On the other hand a driver can get information from other people, e.g., through radio (jamming etc.). This is referred to communication between vehicles or infrastructure (C2C/C2I). These two ways of sensing are indirect possibilities to get additional second-order information about the incoming environment.

2.2 Horizontal layering integrated into the vehicle

The integration of new layers into the vehicle must always be reverse compatible in case of failure and still give the possibility to add features without any modifications on this layer. In such cases, it will be possible to inhibit the controls based on the layer, which is posing difficulties without corrupting the rest of the architecture. It will then become possible to downsize the level of intelligence of the full platform just below this element. This is why the architecture has to be clearly horizontally layered as explained by Brooks in [20] and shown in Fig. 2.3. This concept has also been used in serial vehicle production such as the Vehicle Dynamic Control by Toyota.

When a vehicle is not equipped with this kind of systems, the driver's command is directly sent to the actuators. If layers have to be integrated, the control of some actuators will first have to be done autonomously. Indeed the gear shift has to be automated to give a direct control to the engine, the brakes and the steering unit. Moreover when a differential has been locked, it cannot open autonomously without specific braking torque to compensate its torsion. Therefore, the steering angles will have to be disabled to avoid mechanical defects of the differential and the vehicle will have to stop as quickly as possible.

The first layer named *reactive safety* corresponds to the operational function. It can also integrate the functionalities corresponding to the reactive safety such as braking control, yaw control etc. This layer is based only on the difference between the command and the vehicle state. It will provide a correction if necessary. The connection with the layer below can be either done mechanically by using epicycloids on the steering wheel [71] or a second valve for the baking units. This connection can also be done fully electronically with the integration of a drive-by-wire technology [117].

The second layer named *predictive safety* corresponds to the tactic function. At this level all the possible predictive assistant systems will be inte-

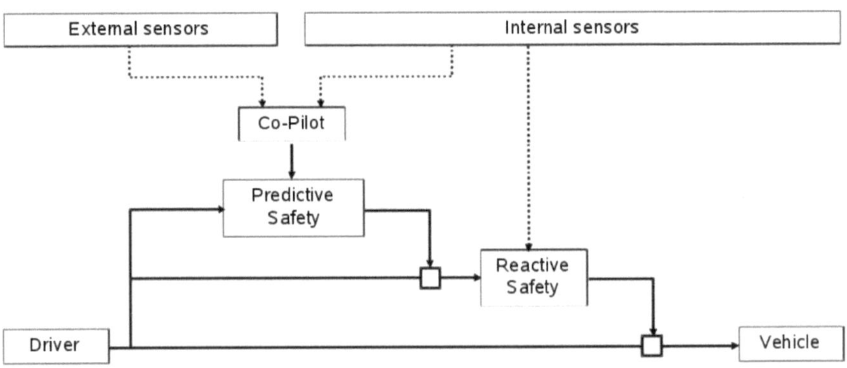

Fig. 2.3. Layers for the vehicle platform

grated in a virtual driver. This deliberative level requires also an access to the driver's command to intervene while an optimization is required. Theoretically three possibilities exist: first a limitation of the driver's command to stay within the safety limits, second a correction or third a decision to realize one of both commands. Since the second method is purely reactive, it can only bring collision mitigation. Thus the innovation will be the combination of the two other methods. Here the connection to the lower level is done electronically such as in [11]. It would be also possible to do it mechanically by adding and moving mechanical limits to the steering wheel and pedals, but the complexity of such devices presents singular systems like the devices such as the German society Paravan.

The strategic function into the cognitive model will not be integrated as an additional layer in this architecture. Actually the strategy function is not relevant for the safety of the vehicle in a short time range; it is more focused on the driver comfort. Moreover the information coming from this layer, typically based on old data of the navigation platform are seldom upgraded (less than 5% for the Blaupunkt navigation platform [22]). Thus the use of such devices as safety input is not appropriate, but it can help the external sensors to perform their tasks. So the connection of these two layers can be a stream delivering the data from the navigation platform.

Moreover an external trigger of the navigation platform is required while danger occurs. Indeed the use of the navigation platform at such moments will disturb the driver, whose brain switches from normal mode to emergency mode and it will skip automatically the strategy function.

2.3 Overall presentation of the new concept

In this chapter a new methodology for the cooperation between the driver and a virtual driver will be presented. The part II defines the new vehicle platform required to design an adaptive virtual driver. After that, this virtual driver is integrated on those platform (part III) and finally the intervention process is developed (part IV).

The innovation of the vehicle platform is the capacity of instability anticipation which enable the assistant systems to pre-compensate hazards. Thus, the use of real-time models of the actuators to extrapolate the vehicle capacities during the next instants will be described in chapter II.4. However, the use of drive-by-wire components will require an energy management to look at the influence of the battery state on the actuators' capacity (section II.4.2). The road–tire friction coefficient will also be estimated (chapter II.2) as it influences directly the single link between the vehicle and the road. Once the vehicle dynamics are modeled, it will be possible on one side to realize the vehicle command with a predictive adaptation (chapter II.5) and on the other side it will be possible to send a feedback to the command level related to the vehicle capacity. On top of that, reactive assistant systems such as ABS and

ESC will be used with a predictive configuration (section. II.5.4) as they may be required anyway.

Once the vehicle dynamics are made available at the command level, the virtual driver begins to bring the tactic level within the vehicle. As the integration of assistant systems reaches an exponential level of complexity, a multi-agent platform (chapter III.1) will be developed to give the possibility to plug these functionalities in parallel. A blackboard will store the outputs of the smart sensors and it will combine them to improve the reliability of the environment models. To improve the level of availability, a redundancy mechanism has been used by integrating algorithms from the robotic field such as potential field method or dynamic window method (chapter III.2) parallel to the event-based algorithms from the automotive world such as the distance control or the lane keeping control (chapter III.3). As they bring different control dynamics, their outputs require a combination on the acceleration–steering space to define the safety envelope of the vehicle command.

When the virtual driver is ready to act in parallel to the driver, its cooperation capacity can be used. In the last part, the different levels of cooperation which are used as degradation modes will be described in details. The first level of cooperation will be purely binary: once the driver's command has reached the safe limit of a maneuver, the system will intervene and align the command with the corresponding safe maneuver. This kind of intervention is only an extension of the emergency brake to the synchronization of longitudinal and lateral directions. Thus, a first required task will be to understand the driver's command in terms of maneuver (chapter IV.2). After that, a transition path can be planned when the intervention kicks in. The confidence values of both driver and virtual driver were not taken into account yet. Thus, another task will be in charge of the driver monitoring to determine his/her drowsiness (chapter IV.3). Hence, a fuzzy intervention can be performed depending on their confidences. Finally the level of assistance required by the driver will be estimated to enhance the cooperation process (chapter IV.4).

2.4 Presentation of the concept of adaptive cooperation

For airplane ergonomics and nuclear power plants, adaptive cooperation has already been investigated since the early 1980s. Norman [95] proposed the idea of cognitive engineering to answer to the increasing complexity of the systems. In 1984, Sheridan defined for the first time the different levels of automation and their inherent risks [111]. From this enumeration, Boy defined 6 years later in [16] the properties of integrated human–machine systems and he/she has shown clearly the performance of cooperation using of real tests in the space shuttle for the landing control. Following this, he/she focused more on the interaction processes [15] and especially how to bring the most relevant information to the human during decision–making process. A derivated concept dedicated on ground vehicles came later from Flemisch in [40]. In his/her approach the trio—human, intelligent assistant system and machine—is fused into one sequential two-level architecture with a driver and an intelligent vehicle. This approach focuses more on the transfer of driving tasks and the ergonomics related to this transfer, by giving the possibility to amplify the driver's request. The limitation of these different approaches comes from their global vision of the acting tasks; and hence they try to provide an intelligent assistant system as one single entity.

Besides taking into account the driver cognition model as dataflow, the approach presented here will also modify the former trio driver, virtual driver and vehicle platform by adding a module involved into the cooperation and the dispatch of its results, see Fig. 2.4. The cooperation will be split independently into strategy, tactic and operational skills. Therefore the driver and virtual driver will have the possibility of bringing their knowledge on different levels only for some specific tasks, for which they get the advantage. For example, the driver gets good capacity for the anticipation and the tactic when he is concentrated; it remains difficult today to have a good anticipation with intelligent systems. On the positive side, assistant systems are fast with repetitive capacity even if the driver has to monitor their actions. For each function, the optimum of the cooperation is continuously evolving depending on their knowledge on the environment. On one hand, if the driver tends to be an expert, the optimal solution will tend to shift generally to his/her side. Even if the environment is well known without risk of unanticipated situations, the assistant system will rarely have to correct the driver's intentions.

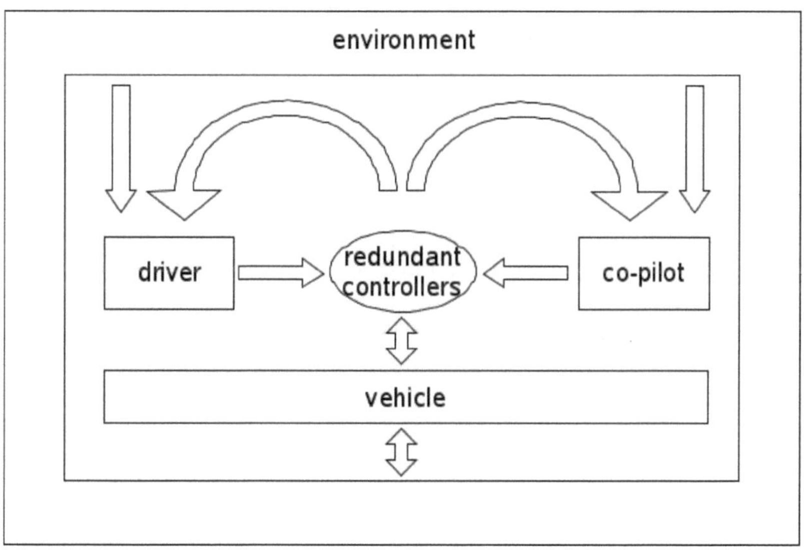

Fig. 2.4. Concept of adaptive cooperation with parallel elements in the command level

On the other hand, the shift to the assistant system can be done for inexperienced drivers or inattentive drivers so that the assistant system can perform mandatory actions without overriding possibilities up to autonomous driving.

The intelligent assistant system has to first understand undoubtedly what the driver goals are, not only in terms of vehicle command but also in terms of destination and maneuver to perform otherwise it does not know how to support him/her. Following this it can determine how to cooperate with the driver. This means it has to determine how much support the driver expects, and how much support it can bring to this person. This difference of capability is also important for the driver, since he has to know how far the assistant system can support him/her. For the functions not related to the short-term control loop, its help is always punctual as events do not come continuously. However, as already explained before, this help may be disabled during dangerous situations to avoid disturbing the driver. For the tactic level, help for the anticipation task is also rarely given and only when the driver will be missing relevant information, such as *what is happening after the driven curve*. For the reactive task, the continuous cooperation requires from the cooperation module to know the driver's acceptance level for the feedback. Depending on the cases, this information will just describe the dangerousness of the maneuver, showing the risky zones for a maneuver change (depending on the driver's own confidence), or describing the source of dangers at a later time. In the worst cases, the assistant system will have to disagree with the maneuver changing and the cooperation module will have to explain to the driver the different reason (e.g. dead angle). Last but not least, the operative level gets a wide range of possibilities, from completely passive assistant system to continuous optimization of the driver's maneuver by means of autonomous driving. The level of delegation may be contested from the cooperation module anyway if the driver's fitness goes down, since the driver may leave the control loop after that. The highest problem here will be to determine the difference between the support requirement and possible support reserve.

Next, if the cooperation is analyzed over the time during crisis resolution [73], three steps will be defined: anticipation, interaction and recovery. The anticipation time is almost the most common level and it corresponds to the case described just before. Potential hazards have to be detected as soon as possible and the dangerousness lowered. An easier interaction for the crisis solving will depend mostly on the human competences and the usability of the interaction process with the assistant system. If their solutions are converging, the resolution will be faster and with less stress for the driver [140]. However, in some cases, their solutions can be conflicting due to lack of information or knowledge for at least one of them. To solve this conflict as fast as possible, the role of the feedbacks to the driver and to the virtual driver are fundamental. Indeed, assistant systems can do mistakes like humans if they would have sensing failure, problems with situation categorization, or control not adapted to the situation. The assistant system can modify its strategy after verification if some elements have not been detected correctly. However, if the outputs stay incompatible, it will be necessary to choose the source of command, which has the highest probability to achieve the recovery, and the other one will be set out of the control loop temporarily. This choice will correspond to a passive recovery if the assistant system has to intervene, or it will correspond to an active recovery if the driver achieves the recovery by himself.

Part II

Executive level as vehicle platform

1

Requirements for the executive level

Abstract. This first chapter is a short introduction into the engineering of the executive level presented in this part. Having defined the tasks of this platform, the choice of the methodology used supporting the modeling of the vehicle capacity will be explained. Then, each part of the data flow will be detailed in dedicated sections.

The vehicle capacity depends on the μ road–tire friction coefficient as it is the only link to the road (chapter 2), on the drive train characteristics (section. 4.1) and finally the influence of the energy state (section 4.2). Once the vehicle dynamics are available, the vehicle platform can perform correctly the vehicle commands (section 5.2). Finally reactive assistant systems will be integrated (section 5.4) as the actuators' capacity may make the vehicle state out of its linear range, e.g., when too much torque is applied (first gear, icy weather, etc.), requiring command corrections.

1.1 Tasks of the executive level

A human driver learns the dynamics of the vehicle over time. However, at each moment, the driver is looking at its behavior to improve the estimation of its current limits. As the driver corresponds to the control unit within the vehicle, he/she is always looking at any feedback to close the control loop. Hence, the driver will increase the safety distance to the vehicle ahead, for example, when the road is icy. Intelligent assistant systems also need to consider the actual vehicle capacity to have the possibility to adapt the driving strategy instantaneously. In dangerous situations, well-trained drivers recognize if the vehicle can perform a critical command to avoid accident or not. In first approximation, the driver can estimate if the vehicle is fast and powerful enough to perform the safe command. Hence, the information that the driver perceives over the time is the latency time of the drive train to reach an extreme state.

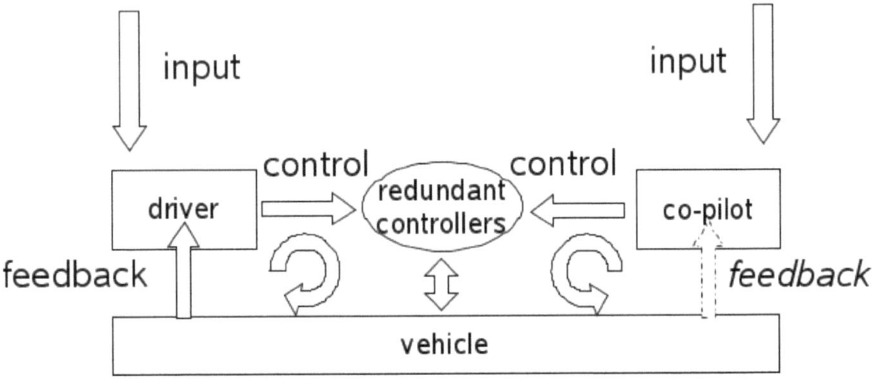

Fig. 1.1. Needs of a feedback for the virtual driver similar to a real driver

Here an approach based on an artificial feeling will be used for the virtual driver in order to go from a subjective adaptation to an objective adaptation. This feedback has to be generated in parallel to the information to the driver as shown in figure 1.1. The executive level manager will first receive information related to the road–tire friction coefficient to estimate the maximal reachable propeller force. Then it will consider the feasible propeller forces that the actuators can perform that has to be adapted depending on the available energy.

1.2 Integration of the predictive vehicle dynamics model

1.2.1 Selection of methodology for the prediction of the vehicle dynamics

Standard models exist to estimate the vehicle dynamics, from simple bicycle models to complex two-track model. Their differential equations can be adapted to describe a wide range of vehicles from a small passenger car to a heavy good vehicle with trailer-combination. With such equations, the estimation of the vehicle states can be extrapolated with different levels of robustness. Rough linear or polynomial extrapolations can be sufficient for short-term extrapolation, but these extrapolations are unfortunately always diverging. Thus differential equations may be integrated as state estimator [141] into a Kalman filter or even an extended Kalman filter to take into account non-linear Pacejka's tire model on top of the two-track vehicle model. The filter horizon can vary after that depending on the needs.

With such estimators, the parameters of the vehicle such as the position of the center of gravity, μ, etc., can be estimated. However, such an approach has two main disadvantages. First, there are actually two filters in one. Out of the first filter the vehicle parameters can be estimated. They are used in the second filter for predictions. Thus the integration on embedded platforms is really difficult. Furthermore, the prediction only deals with the vehicle state and not the maximal limits of the vehicle dynamics.

1.2.2 Integration of the predictive vehicle dynamics model

The approach described above will be based on the self-estimation of each maximal dynamics of the actuators and on the estimator to combine the results into the differential equation.

First a predictive model as transfer function is needed for each unit such as engine, gear, clutch, differential, steering unit, braking units, etc. and secondly the vehicle dynamics model combines these transfer functions. This task is performed in three layers and split into dedicated ECUs as shown in figure 1.2:

- *Estimator*: Estimation of the limiting parameters such as μ, vehicle load, position of the center of gravity, available energy.
- *Units* (engine, gear, clutch, differential, brakes and steering unit): Identification of transfer functions. The challenge lies in finding the optimal compromise between accuracy and software loss.
- *Wheels control*: Combination of the torque estimations coming from the powertrain and the braking unit. The additional reactive controls (ABS, ESP, etc.) will be integrated here too.
- *Platform control*: Computation of the general dynamics model describing the longitudinal and lateral dynamics of the vehicle by using a bicycle model.

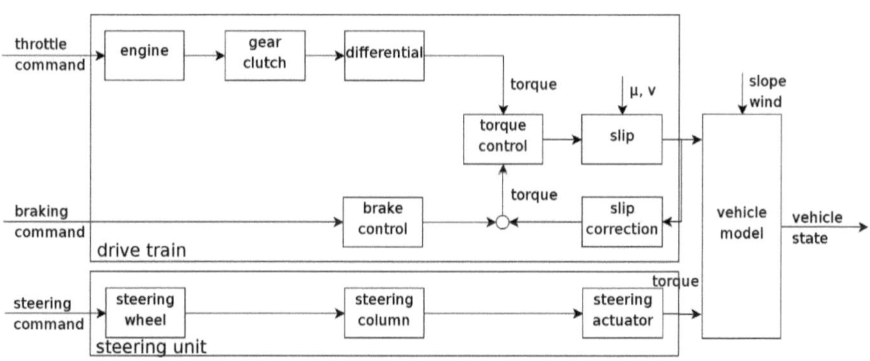

Fig. 1.2. Functions integrated into the vehicle platform

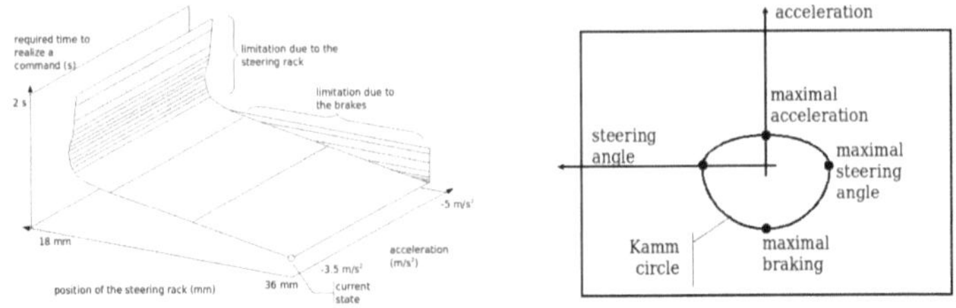

(a) Realization time of each vehicle command

(b) Extrapolation of the limits by use of the Kamm circle

Fig. 1.3. Fusion of the envelope transfer functions by using the Kamm circle

At the end, the model provides the maximal acceleration, minimal acceleration considering engine brake, minimal acceleration with brakes intervention and for maximal steering capacity. The transfer function describing the minimal acceleration with engine brake will not be sent to the command level, but instead will be used to select the best braking parameters.

As these transfer functions are taken independently for single acceleration, braking or extreme lateral acceleration, they have to be integrated together by taking into account also the tire dynamics. Thus the four extremities of the Kamm circle [66] for the maximal acceleration, minimal acceleration and the extreme lateral acceleration will be defined with the transfer functions. They are presented in figure 1.3(a) as dots with the Kamm circle. With a calibration of the tires' model, it will be possible to extrapolate the limits involving both acceleration and steering.

In figure 1.3(a), the transfer functions for braking and steering are overlapped. The resulting maximum delay between both transitions gives a 2D graph which provides the transition time to each command as acceleration and steering. In usual cases, the steering will be the slowest actuator. That is why it will set the transition time to most commands. This is illustrated in figure 1.3(b). But for hard braking harder than -7 m/s^2 and corrections smaller than 22 mm, the braking will take longer, which explains why they set the transition time in this range.

2

Road–tire μ friction coefficient estimation

Abstract. This chapter will focus on the estimation of the μ-coefficient. This information will be used in another chapter to determine the maximal force which can be transformed into kinetic energy. After a short discussion on the current methods of measurement, a new approach based on predictive and local estimations will be explained.

The driver estimates the μ-friction coefficient through the recognition of the road surface condition and the quality of the contact road–tire. The measurement of the road surface includes sounds, vibrations, colors, shapes and surface gloss. Similarly to humans, every vehicle could use such inputs to deduce the μ-friction. The measurement of the road surface could be directly done with some external sensors such as camera, laser or microphone or with internal information such as the wheel rotation speed.

Additionally to the different direct sensing methods, the driver combines the different results to improve the accuracy of the measurements or to detect a wrong assumption coming from one sensing channel. This second part of the methodology needs additional information about the environment such as weather, time of the day etc. This indirect estimation is not accurate enough to provide a reliable estimation by itself but it is supportive of checking the reliability of the estimation.

2.1 Analysis of the current methodologies

Several methods to estimate μ are based on the vehicle state: internal sensors are used by an auxiliary function conjointly with a dynamic vehicle model. The actual state of the vehicle and the reaction to a motion command allows to extrapolating the limitation of the realization of the given command. The single link between the vehicle actuators and the road is the contact road - tire. Therefore it is possible to estimate this friction coefficient. Three different approaches based on this idea are already under investigation as reported by

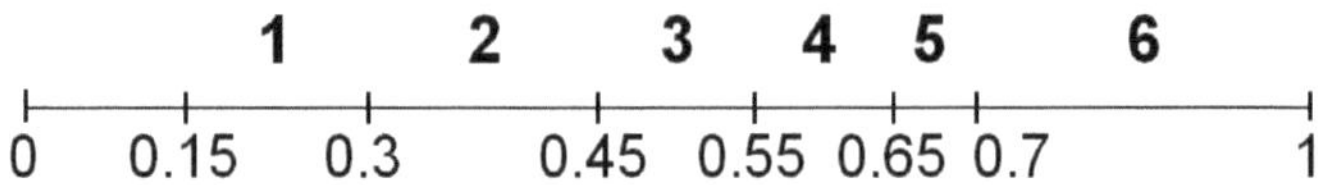

Fig. 2.1. Ranges of μ

Wang [132]. Each approach focuses on a different level of the vehicle-road model to extract some rules about the road-tire friction.

A vehicle-based system [53] and a wheel-based system [51] determine the slip value and use a tire dynamic model to extract the μ-friction. The relation between the slip and μ is extracted from the Pacejka magic formula [98]. The single difference between both methodologies lies in the way to determine the slip value. The vehicle-based method uses measures of both lateral and longitudinal motion, and wheel-based approach extracts the torque transferred to the wheels from the powertrain or/and from the brakes.

A third method is based on measurement of tire deflections from piezoelectric strips embedded in the regular tires. The piezoelectric strips can measure local deflections of the tire. As the embedded piezoelectric moves through the contact patch of the tire, the deflections in the contact patch are used to calculate the forces on the tire as well as the longitudinal slip and the slip angle of the tire. A tire model is then used to calculate the tire–road friction. The inherent problem with this method comes from the discrete measurement on the tires and give data with too low frequency. Furthermore, the integration of several sensors and their wireless communication to the body shield implies high cost for serial implementation.

These four methods of measurement are dealing with both road surface and tire. Therefore their accuracy can be relative good but can only provide past measures. In the next couple of years the assistant systems will have to deal with a wider environment. Therefore exterioceptive sensors such as camera and laser scanner will be needed to estimate the μ coefficient ahead the vehicle.

Unfortunately, as they cannot deal with the tire state, they can only deliver a worth case estimation. Therefore the predictive estimations are up to now depending on a calibration of the tires' model contrary to the auto-regressive measurements, which could determine the characteristics of tires.

With these predictive methods, six different classes of μ can be defined in the same way as in [29] and shown below: (1) snowy, (2) wet, (3) humid cobblestone, (4–6) from bad to excellent coefficient (Fig. 2.1).

2.2 Predictive camera-based measurement

The main aim here is to determine the type of surface conditions ahead the vehicle: road surface, humidity, snow etc. Three different approaches are possible for the analysis. Each method analyzes the images with a different level of granularity, from global measures down to local measures. These will be shortly described before improvements are proposed.

- Statistical approach: the luminance of the pixels and their repartition on histograms are analyzed to extract some properties. This approach takes the image as a global source of information. Thus the signal processing is to filter a huge amount of information. On the other hand, the robustness of the processing amongst local defect is interesting.
- Macroscopic approach: local variations are analyzed to match patterns of standard road surface structure. This method will determine the type of road texture and the humidity grade. It will provide the most homogenous concept as it will have the robustness of a global approach, but it will also look at local details such as the microscopic approach described next.
- Microscopic approach [76]: a two-dimensional Fourier transformation is applied on the picture to model the irregularities of the road and the regular patterns. With quality images, this method will enable to determine locally the road topology and the humidity–snow grade. Even if it can provide the best results, the robustness to local defect will be less accurate.

As Kuno and Sugiura explained in [75], a robust use of the statistical approach needs a learning system. Due to the needs of computation capacity, it cannot be integrated into embedded systems yet. The last approach based on the high frequencies of variation of road texture is quite difficult with the use of a camera, which is continuously vibrating due to limited stiffness of standard housing. Therefore this approach will be difficult for moving vehicle if the camera is not attached with a special device to compensate the low pass filtering of the vibrations.

To simplify the hypotheses, Kuno and Sugiura supposed in the same article [75] that μ is only depending on the luminance of the road surface: a gloss surface describes high humidity on the road and thus a low μ. They extract the luminance to generate its histogram over the image, and they combine that information with a look-up table to estimate the friction coefficient.

However, the road pavement should be considered to improve the estimation. This is why their limiting assumption will not be considered here. To take into account the road pavement, a statistical approach (section 2.2.2 of this chapter) has been used to extract some patterns of the road surfaces. The results of this statistical approach are transmitted to a macroscopic approach (section 2.2.3 of this chapter) by using a mathematical model. The macroscopic approach will be used by integrating both texture analysis and gloss measurement together. Actually this approach will be partially similar to estimation of landscape characteristics based on snapshots taken by satellites as reported in [7],[23], or [100]. Some properties of this statistics containing variables such as the global luminance, the energy or the variance can be used to characterize some types of textures.

The macroscopic approach developed here requires three computation steps. The co-occurrence matrix from the different ranges will be calculated, and hyper-spheres will be matched on this matrix. Next the hyper-spheres will be compared to the different models of the current luminance. Finally the prediction of μ over the distance will be correlated with the other past predictions to enhance the reliability by detecting errors due to changing lighting, for example.

2.2.1 Extraction of the ranges analysis

The results of lane detection such as in [9] define the lateral limits of the range to analyze. In addition an algorithm for detection of objects defines the maximal distance ahead the vehicle defining the visible road section. Then a bird view of this road section in range is computed depending only on the lateral intrinsic parameters of the camera (contrary to other full methods as in [18]). The other corrections are not needed for the depth of the image because the picture will be split into several small areas (see figure 2.2) for local analysis and the residual errors will not be significant enough to be taken into account. To simplify the algorithms, a static height of 30 pixels is set for the different areas. Up to nine different areas to monitor will be on the (640 x 480) picture. If the areas are not wide enough, the results of the computation will not be accurate enough as some important variation of luminance may appear occasionally.

2.2.2 Statistical approach

The statistical approach is based only on the picture histogram analysis and it attempts to identify standard patterns. This approach should be feasible because it requires less information and data mining than an approach dealing directly with pattern recognition on the image itself. In fact, this second

Fig. 2.2. Split of the range to monitor

method requires more computation to be reliable for each type and size of cobblestones, for example. In normal cases without CCD saturation, specific histograms can be obtained from the snapshots, as shown in the figure 2.3. The abscissa defines the possible levels of luminance (between 0 and 255) and the ordinate defines the number of pixels with such a luminance corresponding to a luminance bin.

The snow can be easily detected because the luminance range of its histogram is not corresponding to the others as it is centered on the high luminance (effect of the white color). Normally ice can be detected through the same patterns but differences between snow and ice are difficult to distinguish. Furthermore the sand can also be detected because the histogram ranges of the different colors are not overlapped (sand has no neutral–grey color). Last but not least, cobblestones can be detected most of the time because of their asymmetrical histogram. From a concrete point of view, a model is computed

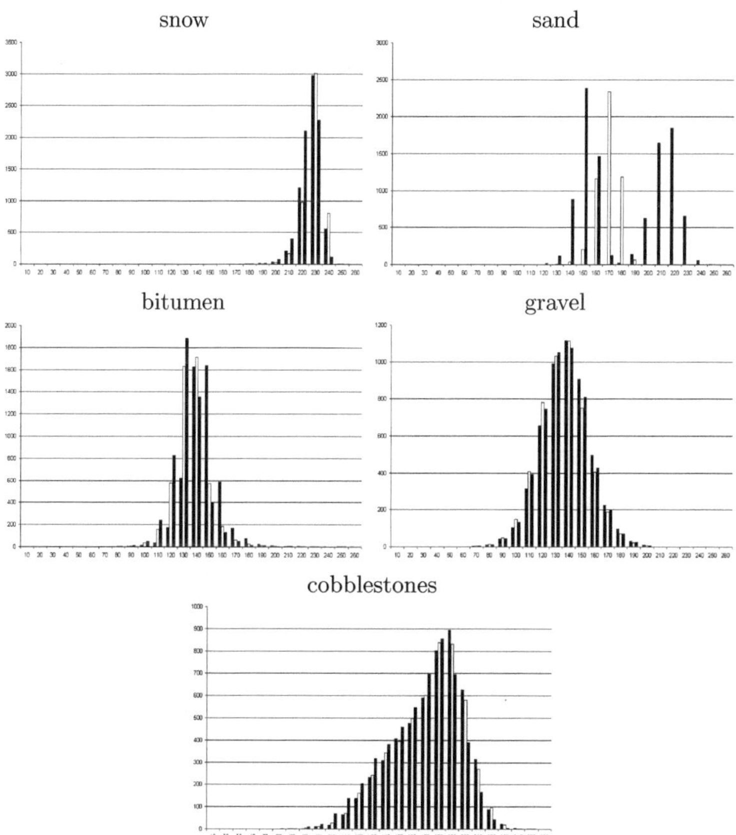

Fig. 2.3. Samples of road and histograms of luminance distribution at pixel level

for the right side and left side of the maximum of the histogram based on least square method. After that, the normal vectors of the two lines will be compared.

However, the estimation is more challenging for the other surfaces related to associated to rubble with different granularity: bitumen, and all variations of gravel. In those cases the luminosity and the humidity strongly influence the mean and the shape of the histograms. Luminance and humidity can transform the histogram of dry bitumen in the shadow similar with the histogram of wet gravels. The figure 2.4 presents such similar histograms, which are difficult to classify without additional information. Although those surfaces have the same luminance distribution, they do not have the same properties. Therefore, it is difficult to estimate—correctly and with satisfactory robustness - the friction coefficient without using another input.

Hence, a second input has to be extracted from the histogram of the derivated picture, which can be obtained correctly from a Sobel filter. These histograms can be approximated as Gaussian curves, as seen in the figure 2.5. The luminance influences slightly the position of the mean value while the humidity acts on the height of the maximum as it affects directly the gloss of the road. Additionally, the roughness of the road surface is correlated to the width of the Gaussian curve.

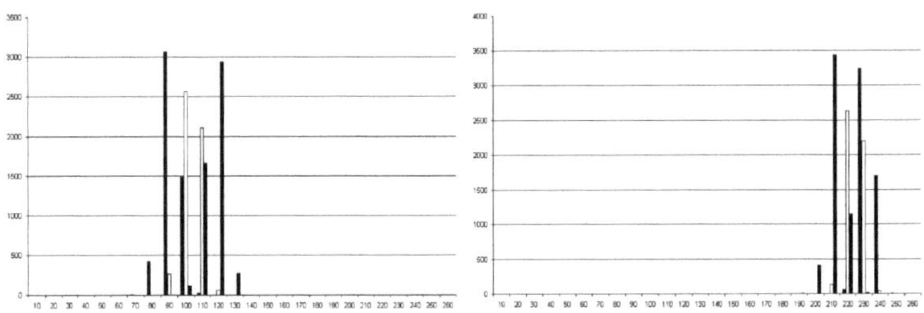

(a) Histogram for a given road section with shadow or not

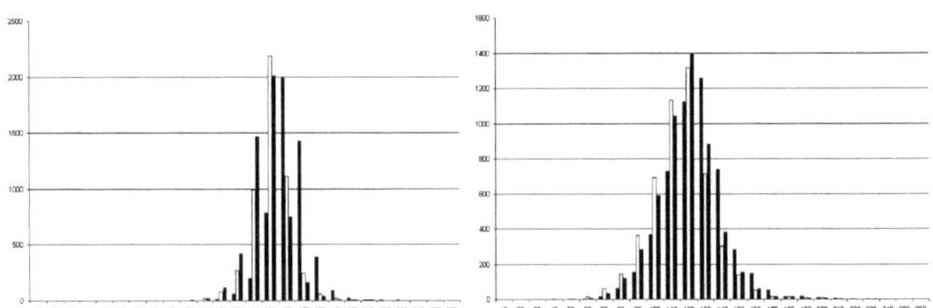

(b) Histogram for a given road section with wet or dry conditions

Fig. 2.4. Action of the luminance and humidity on the histograms of luminance

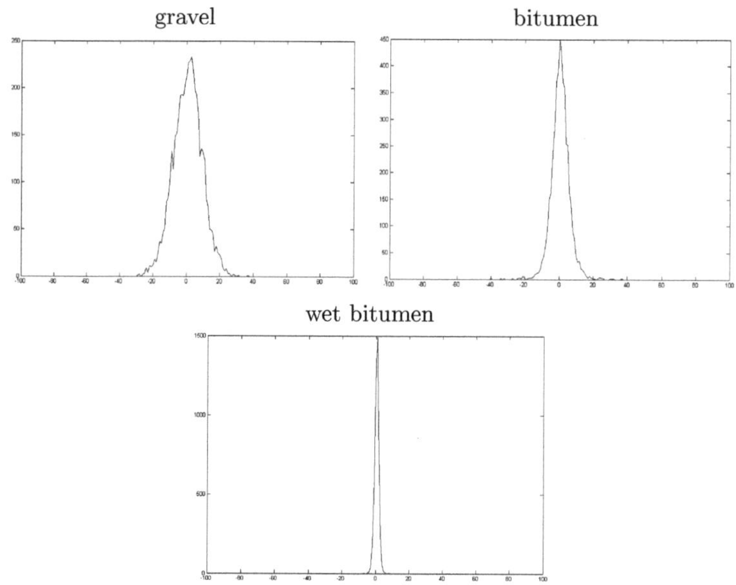

Fig. 2.5. Histograms of the derivative of the pictures defining the amount of pixels with the same gradient of luminance

After having analyzed all the videos taken, it has been possible to determine correctly the size of the histogram ranges for the derivated pictures. Hence a classification system is able to estimate in most cases the friction coefficient thanks to this additional information. However, the amount of parameters for the pattern classification have to be reduced for integration in an embedded platform. Therefore, a second approach dealing only with the histogram and the width of the Gaussian of the derivative will be used.

2.2.3 Macroscopic approach

This section will present the design of the off-line patterns and the algorithm performed online to generate the adequate patterns. The probabilistic pattern recognition will be described after that, and finally the μ-graph verification will be explained.

Co-occurrence matrices

For a picture I with n levels of gray (here 50), its co-occurrence matrix M_I with the dimensions $n \times n$ will describe the amount of pair wise of pixel luminance on the picture. Let take each pixel $p_{i,j}$ with the index (i,j) and luminance k, and its right neighbor $q_{i,j+1}$ with a luminance l. Let us take also the co-occurrence matrix with a size of $n \times n$, which is set to zero at the beginning. For each arrangement (k,l), the value of the cell with this index

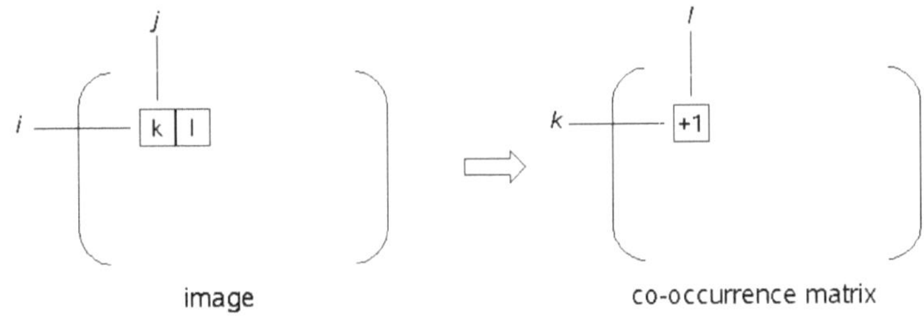

Fig. 2.6. Example how to feed the co-occurrence matrix

will be incremented. For each co-occurrence matrix shown here after, the axle will be automatically corresponding of the indexes of luminance and the size is 50 × 50 (Fig. 2.6).

$$\forall(k,l) \in [1,50]^2, M_I(k,l) = Card_{i,j} I(P_{i,j}) = k \& I(Q_{i,j+1}) = l \qquad (2.1)$$

Two properties of this co-occurrence matrix will be interesting. First of all, the sum of the values for one row is defining the amount of pixels with the same luminance. Thus, the shape of the Gaussian curve defined statistically before will be found here again on the row axle. The second property is dealing with the trace. Let us take once again the two former pixels p and q. For each pixel within the co-occurrence matrix on the index (k,l), the luminance difference with the right neighbor is $l-k$. Thus the gradient of luminance is defined by the co-occurrence matrix by the distance to the diagonal. Hence, the Gaussian curve defined statistically before can be found here again by looking at the summation of cells along the diagonal.

The difference between both matrices can be used to estimate the type of texture. A three-dimensional rendering of these matrices looks like a Gaussian surface. Actually the position of the maximum on the axe of symmetry and the spread over this diagonal are related to the histogram of the picture. Furthermore, the width of the form is related to the size of the histogram for the derivate picture. Consequently all the inputs defined with the statistical approach are also emerging here altogether. Its maximum will be used as first characteristics because it corresponds to the averaged luminance. The spread of the surface along the axe of symmetry and the wise perpendicularly to it will be the other characteristics. Here the height of the maximum will not be taken into account, even if it describes indirectly the homogeneity of the picture. Actually this information can be found indirectly within the co-occurrence matrix by having a look at the distance between the different extremes. Moreover, Gaussian surfaces will not be used directly to accelerate the computation but only the definition of their limits (lozenge on Fig. 2.7).

The form of these curves will be well defined for the road sections close to the vehicle as shown in figure 2.8, but the farthest sections will get noisy matrices, because of the distance and the image correction. Therefore their

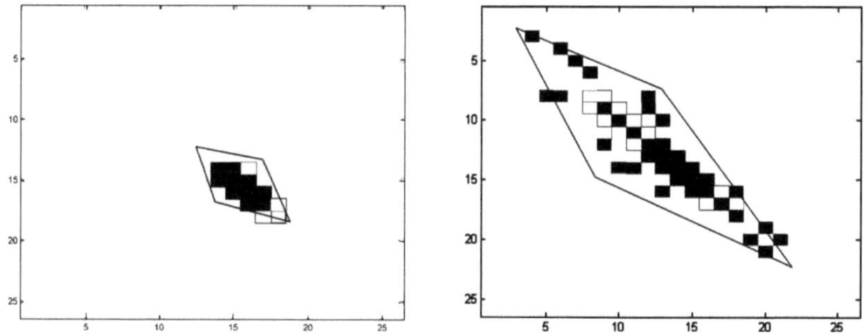

(a) Macadamized road (b) Cobblestones

Fig. 2.7. Example of two co-occurrence matrices

corresponding matrices will tend to go to the white levels (bottom right) by good luminance or to the dark levels (up left) by night. The closest section (top left in the figure 2.9) gets the best accuracy for the estimation of the friction coefficient. But during the night only the ranges overlapped with the beams can be really analyzed.

Real-time extraction of the envelope model

The creation of the patterns is the crucial step for the quality of the estimation because biased patterns will reduce their cognition. All the tests have been performed on the same test track with the same vehicle where different road

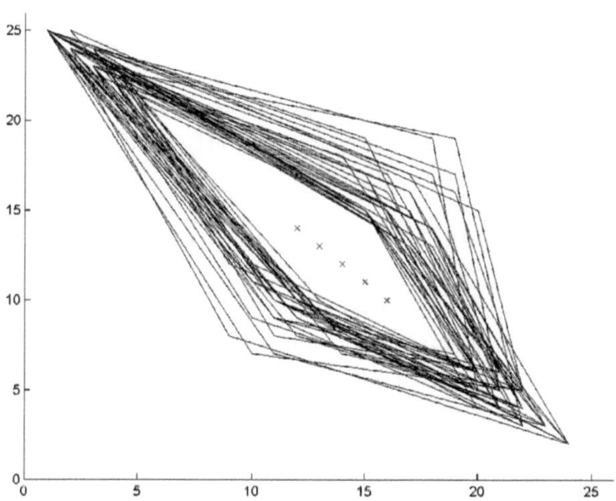

Fig. 2.8. Co-occurrence matrices for cobblestone roads (μ=0.45). Tests performed on the test track by Mercedes-Benz for 1.1 km

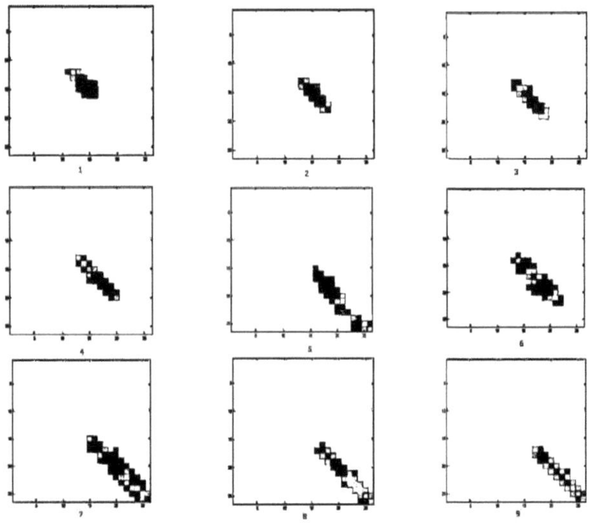

Fig. 2.9. Measured co-occurrence matrices for macadamized roads (μ=0.65). The matrices' indexes are starting from the first range close to the vehicle

surfaces are present. In practice the 1.1km long test track has been driven about 15 times for each kind of weather (snowy, wet, humid or dry) and differences of luminance (time and sun orientation). Figure 2.12(a) shows an example of the image taken during these tests. For each image taken, the image of the road ahead the vehicle has been split into several small ranges on which the co-occurrence matrices have been calculated. The example in figure 2.8 shows the results of the co-occurrence matrices for an equivalent image as the one in figure 2.12(a). Their indexes are defined increasingly starting from the range 1 just ahead the vehicle and the other ones step by step.

Here the influence of the blur (due to vibration, for example) on the image can be seen as the rhombus will move to the bottom-right corner over the distance. The blurring induced by vibration leads to a softening of the luminance gradients. The effect increases as the range of the road sections increases. But this phenomenon will differ during the night as the halo will be dark. Hence the last ranges will have their co-occurrence matrices tending to the opposite corner (dark with pixel intensity about 0).

All the co-occurrences matrices corresponding to a same situation will be matched together. Figure 2.9 is one example of the results obtained for about 500 m on a cobblestone road. In this example, few variations can be seen on the measures. Next, the preparation of the pattern has been realized in two steps. First an average of the co-occurrence matrices has been calculated to provide a statistical distribution of rhombus' limits. Then measures, which are not within the range defining 90% of the measures, will be taken out because they are considered as false or altered measures. Finally the statistical repartition will be computed once again. Its maximum will be used as pattern extremity and the area as uncertainty range.

Hence patterns for co-occurrence matrices of the different ranges have been calibrated for different μ-coefficient and the two extreme levels of luminosity. But these four steps are not enough to cover the luminance range. There are two possibilities, either a pattern is created for each luminance, that requires a huge amount of data, or extrapolation is used to generate a pseudo-pattern instantaneously as shown in the figure 2.10. Actually the differences between the two patterns corresponding to the two extreme levels of luminance (in bold in the same figure) are almost linear with an error of positioning about 8% (corresponding to less than 4 pixels). Hence for a given level of luminance a pattern can be defined between these two extreme patterns as shown in the same figure. To do this, only, the extreme patterns are used (shown in the figures 2.11(a) and 2.11(b)) with the current luminance level. With this approximation, the patterns can be extracted in real time by looking at the position of the maximum of the co-occurrence matrix on the diagonal. Its distance to the two extremes is used to weight their patterns and to determine the adapted one.

Single determination of the graph of μ

After having extracted the patterns on the co-occurrence matrix for the different μ, they have to be matched to the measure from current tire friction model to estimate the preponderant μ-class. Each pattern has to be matched on the rhombus and the distances between the j extremities E_j are measured. Parallel to the generation of the patterns, the standard deviations have also been calculated to determine the probabilities of belonging. The confidence values

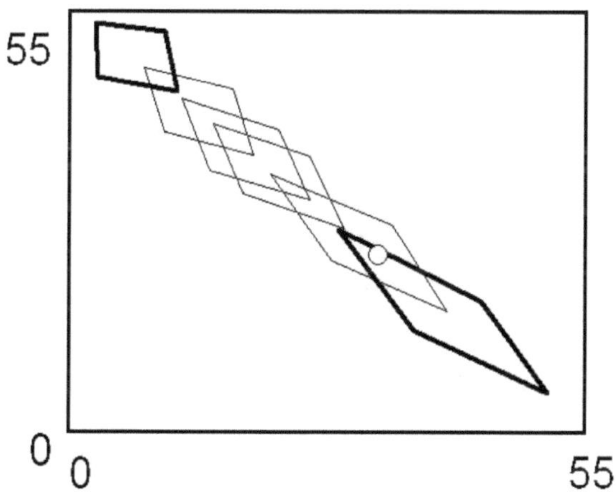

Fig. 2.10. Extrapolation between the two extreme patterns

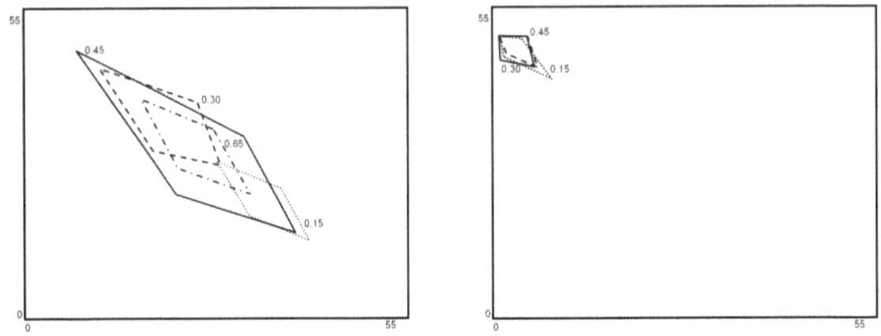

(a) Patterns for maximal luminance

(b) Patterns for night purpose

Fig. 2.11. Extreme patterns used for the surface estimation on the first road segment

for the j extremities $E^{\star}_{i,j}$ can be extracted from these empirical curves and the combination of these four inputs set the probabilities P_i of the i different classes of μ with the equation 2.2.

$$P_i = \frac{\prod_j a_j \cdot P(E^{\star}_{i,j})}{\prod_j a_j \cdot P(E^{\star}_{i,j}) + \prod_j a_j \cdot \overline{P(E^{\star}_{i,j})}} \tag{2.2}$$

The a_j parameters weight the influence of each extremity. In practice, the extremities along the top left to bottom right diagonal have a higher weight because their wide range is easier to separate into different classes. The estimation of μ is calculated from the combination of the different probabilities by using the next equation. The final class of the friction coefficient is given after rounding the results.

$$\mu = \sum_i P_i \cdot \mu_i \tag{2.3}$$

In the example shown in figure 2.12(b), the graph of μ ahead is quite stable around 0.55 (real value), which is well corresponding to the road (gravel). But two degradations down to 0.3 (humidity) can be seen for the areas 5 (6m) and 8 (15m), they are corresponding to the parts with a high lighting. Here the systems thought it has seen wet parts of the road. Moreover, the maximal stability range is about 70 m in practice, which corresponds to the ranges below 12.

Time correlation of the measures

At this point it is possible to get an instantaneous estimation of μ ahead of the vehicle. The confidence of the measure can be improved dynamically by correlating the different measures and detecting inconsistencies. First the graph of μ needs to be transformed to give μ for all points ahead of the vehicle.

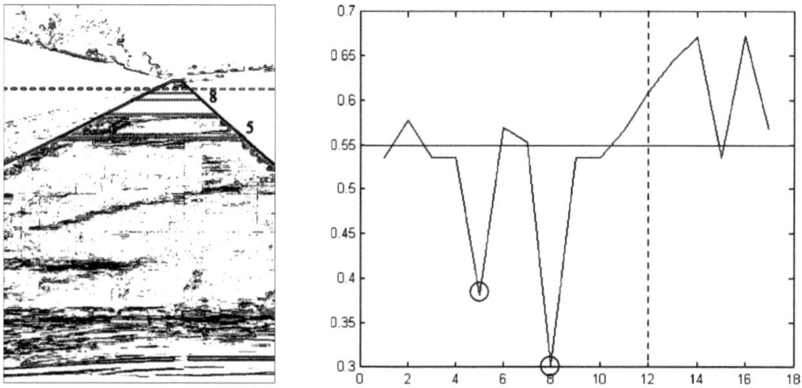

(a) Standard road on test tracks

(b) Corresponding graph of the estimation of μ ahead the vehicle

Fig. 2.12. Example of μ-estimation

As the estimation is made only for regular areas on the picture, this graph will be constant with increasing width of the steps as shown in figure 2.13.

Then the speed needs to be taken into account to match the new measures over the others as in the figure 2.13. The near range corresponds to the road section for which most estimations are available. But all the measures have not the same confidence. Actually the first measures, which have been taken as the vehicle was far from this point, have low weight due to the blur. At 70 km/h about 50 measures can be matched on a same point for a visible analysis distance of 60 m, which is the limit of stable visibility.

That is why the overlapping algorithm needs to take into account a weight, defined directly as the inverse of the distance, for each measure depending on

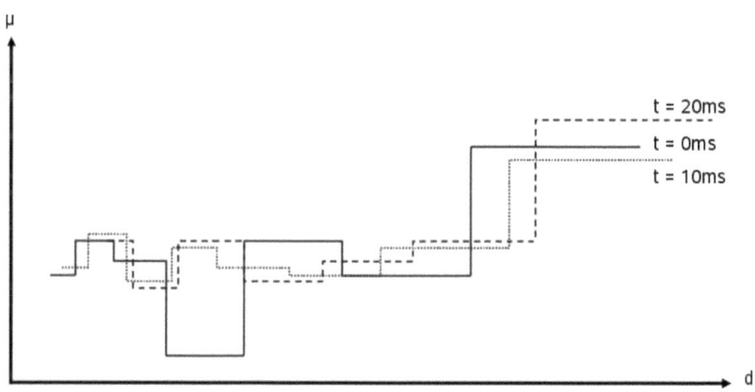

Fig. 2.13. Overlapping of the different predictive measures

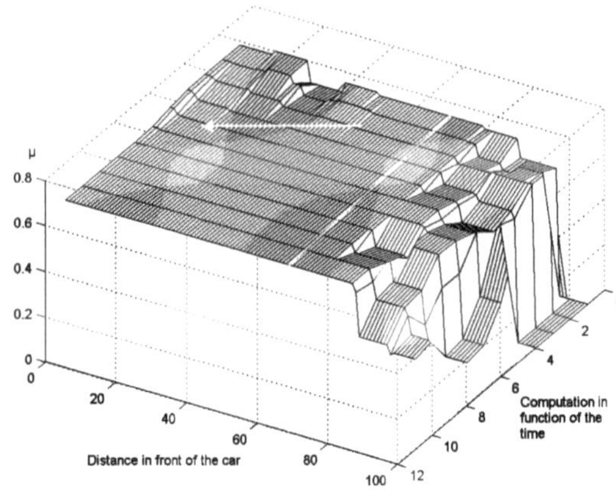

Fig. 2.14. Evolution of the predictive measures over the time by using a time correlation

the time when this measurement was performed in order to define the weighted mean class. Therefore the measures made for a same point are sorted in clusters, maximally one per μ classes. Then these clusters can be weighted to select the most significant class. Finally for this class and the two neighbored classes the mean value is calculated taking into account once again the estimation weight (Fig. 2.14).

2.3 Local microphone-based measurement

After having estimated the μ-coefficient ahead of the vehicle another estimation method will be used, it will be correlated to another way in order to have redundant measures. A microphone will be used here to measure the response of the loud-speaker effect of the road. To do this patterns will be defined offline and online pattern recognition will be done to estimate the μ-class.

2.3.1 Measuring the loud-speaker effect of the tire

Part of the air in front of the tire will be confined between the road and the tire surfaces on the micro-holes, when the tire moves forward. When this air is released after the tire rotation, it hurls out of these holes behind the tire under high pressure. The loud-speaker effect coming from the road and the tire surfaces will generate a characteristic sound (Fig. 2.15).

The sound generated by this phenomenon depends mostly on the pressure between the vehicle and the ground, the nature of the tires for the absorption capacity, the tangential and radial force on the wheel (for the direction of the loud-speaker effect), the speed and the road surface. After tests with different tires, coherent results have been demonstrated. Thus one single general model

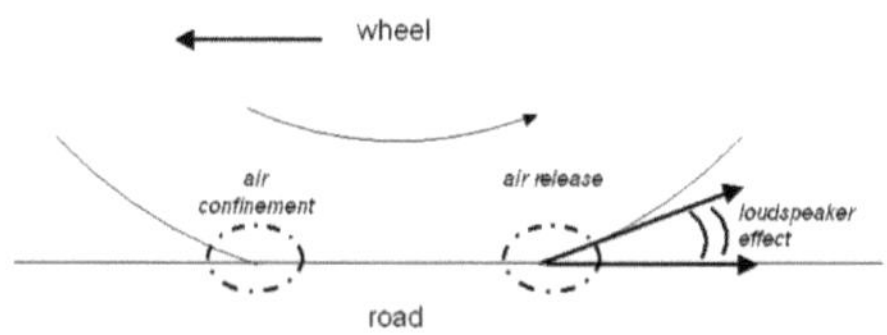

Fig. 2.15. Loudspeaker effect

can be developed considering only the road surface and the speed without taking into account the wheel orientation (Fig. 2.16).

First of all the noise generated by the engine prevents the extraction of the characteristic frequencies as shown in the figure 2.17(b). Furthermore, the loud-speaker effect cannot be measured with the microphone at low speed. Thus the methodology can only work in one defined speed range. Moreover some peaks can be seen on the Fourier analysis for the measures done on the front axle, which cannot be seen for the rear axle. Hence the noise generated by the engine can cover the characteristics sound from the road, if the microphone is too close from the engine. That is why the microphone has always been used on the opposite axle to the engine.

2.3.2 Frequencies extraction from the collected data

The microphone used here is a mono-directional microphone with a bandwidth of 2 kHz. The acoustic frequency range is below 1kHz, where most of the emerging peaks can be extracted.

The speed influences the noise on the measures because of the shuffle, for example. In figure 2.18, three examples of the noise behavior are depicted for

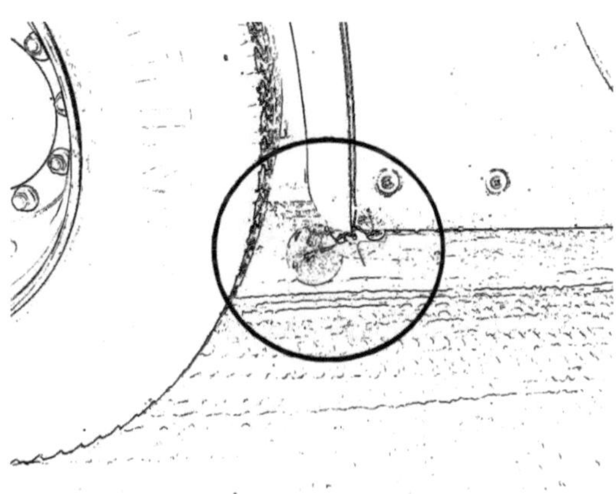

Fig. 2.16. Position used for the microphone to hear correctly the loud-speaker effect

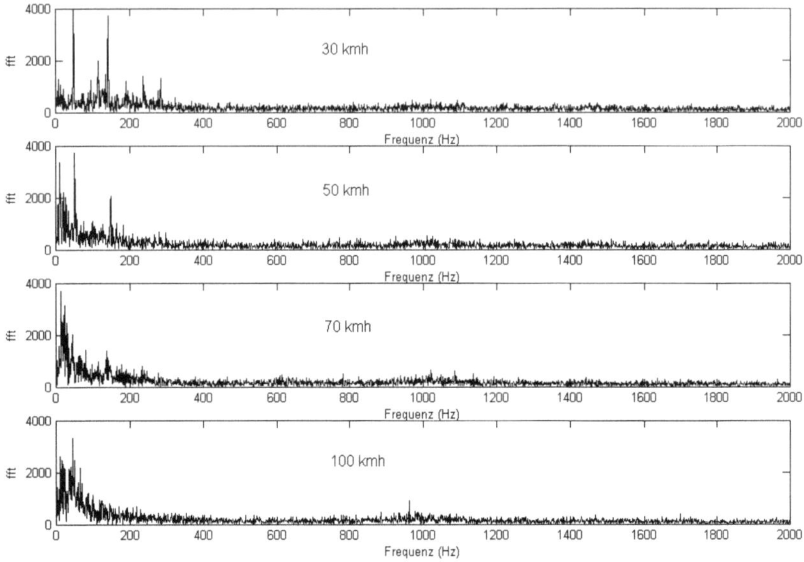

(a) Measures done on the front axle

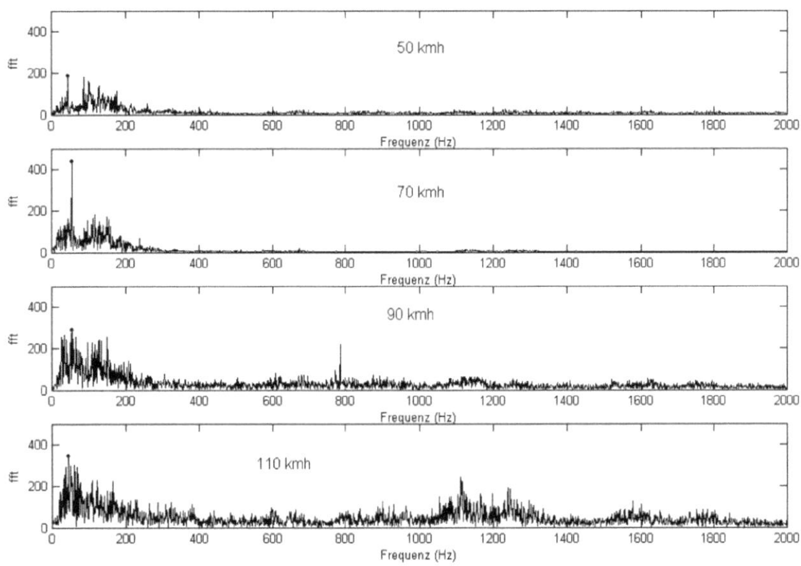

(b) Measures done on the rear axle

Fig. 2.17. Difference of the spectral analysis with a constant speed of 1400 rpm between measures close to the engine (rear axle) and far away from the engine (front axle) the same road condition but not the same speed. The alteration of the signal due to the noise is not relevant for the analysis since it is possible to extract the first characteristics frequencies for speed higher than 30 km/h.

The main problem here is the split between characteristic sounds from the engine or from the road condition. A reliable estimation of μ requires first a good model of the characteristic sound of the engine, which depends mostly on the rotation speed, the gear, the load etc. Let us take the following work hypothesis. First the measures are always performed at constant throttle position, hence the engine sounds is stable. Without these constraints, models will be required for each speed, each gear, each torque request and each μ value. Thus the amount of data would make the system too complex for an embedded integration. Secondly, the engine rotation speed corresponds to the vehicle speed. The system will not have to take into account some special cases where an abnormal gear has been shifted.

The extractions of the frequencies from the engine side and from the road condition side are done at the same time. The frequencies of the peaks are searched after a filtering and they will be stored as potentially interesting frequencies. In parallel, sound has been measured prior to that for different speeds and different road conditions. All the data for a given speed but with different road condition will be subtracted during the first step. Thus, only the frequencies coming from the engine will be visible. Hence, for these data the frequencies coming from the road condition can be extracted as the rest of the emerging frequencies. At the end, a database has been defined in which the frequencies are defined for each vehicle speed. In figure 2.19 the positions of the relevant frequencies are drown for the current speed and for different friction coefficient.

2.3.3 Construction of models

The acoustic signals exhibit a stable power distribution when the road surface stays constant. Moreover some continuous transitions exist for the surface with gravel and bitumen, because the sizes of the peaks are just varying. On the other side, no continuity exists when the surface changes to cobblestones or snow. Therefore, patterns need to be defined for the standard road surface and the estimation can be performed by matching them on the measures.

Measures with the microphone have been realized for several speeds and a set of characteristic friction coefficients. Results in figure 2.19 show the possible regions of interest for relevant frequencies. For each one, ranges have been determined for the frequency itself and for its amplitude. Center positions and variances will be used to define the probabilistic range of the frequencies. Generally, the width of these ranges is about 30 Hz, and this wide range gives the possibility to have an easy research of the interesting frequencies as explained later.

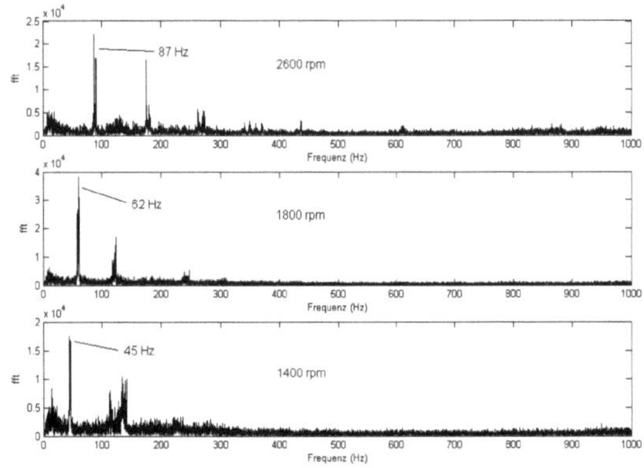

(a) Different engine speed ranges at 30 km/h with a μ of 0.65. The loud-speaker effect can be caught by the microphone easily.

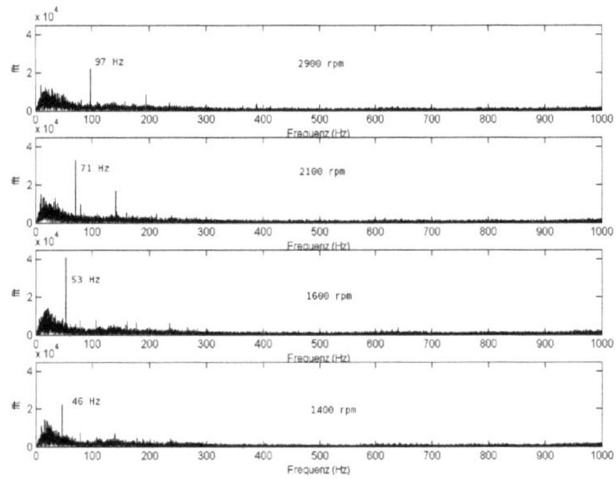

(b) Different engine speed ranges at 75 km/h with a μ of 0.65. The level of noise is important but do not disturb the analysis.

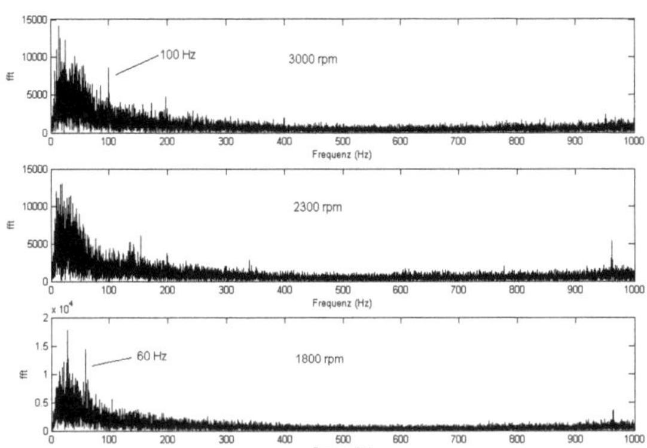

(c) Different engine speed ranges at 100 km/h with a μ-split of 0.65. The noise prevents now the frequencies' extraction.

Fig. 2.18. Sound measured for different engine rpm

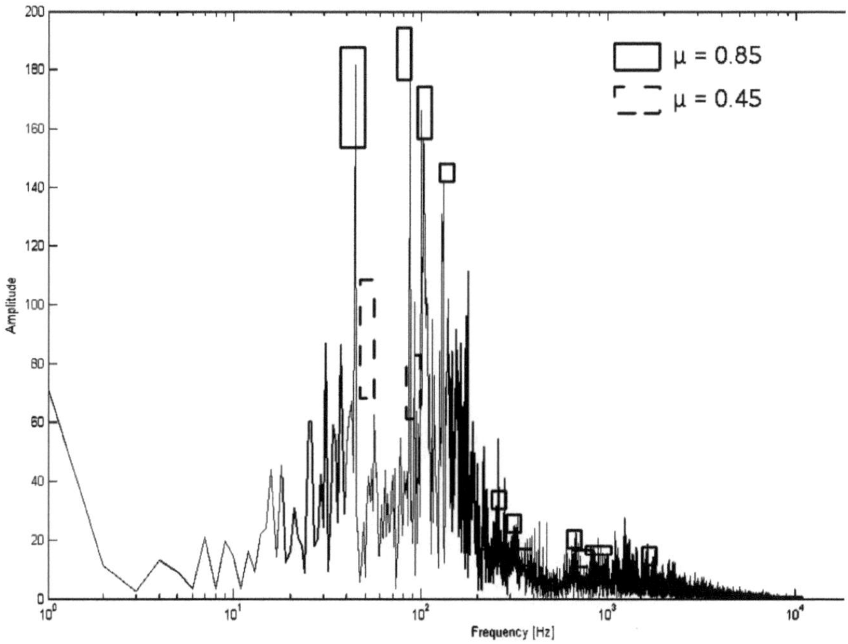

Fig. 2.19. Regions of interest for μ=0.45 and μ=0.85 at 90 km/h

At the end, a small map has been defined where the number of characteristics frequencies and their characteristics are defined for several speeds (40–130 km/h with a 10 km/h step) and for different μ (0.2, 0.45, 0.65, 0.85). In figure 2.19, the possible ranges of the frequencies and their amplitudes are plot depending on the current speed. For this example corresponding well to μ in the class 0.85, almost all the relevant peaks are included in the corresponding ranges. But it may happen for different speeds that some ranges overlap. In these cases, the ambiguity can only be resolved by considering several frequencies.

2.3.4 Matching of the measures

Unfortunately, the number of detected peaks can vary over the time depending on the signal/noise ratio. Typically, the three first harmonies can be extracted without difficulty, and in addition, two other peaks can be also detected in good conditions. The number of peaks defining the pattern is also depending on both speed and friction coefficient. Consequently, the detected frequencies inferred on the models. First the difference between each detected peaks and the two closest peaks in the model are stored. After that the combination giving the lowest global difference will be chosen as matching optimum. After that, the matching can be anyway altered to improve the correlation when frequencies appear or disappear. In practice, such events happen seldom. The allocation of the peaks on the model will start from the low frequencies (as highest influence on the final sound) and no possible misunderstanding is possible until the third peak has to be attributed.

For each peak $\mathcal{P}_j$, the probability in the Fourier space to belong to the model i can be defined as a two-dimensional Gaussian. In figure 2.19, for each interesting frequency a two-dimensional range of presence can be defined as a rectangle. In practice the limits of the frequency range have been defined to include 90% of the measures performed, but no probability will be extracted from the position of the current frequency because the ranges overlaps themselves too much. On the other side, the range of the amplitude will be used to set the probability of belonging. If a peak j of amplitude $A_j^\star$ is included in the range of the 90% of membership, the mean value A_j of the range will be used with the variance σ_j. The probability of the peak to match the model will be defined by the next equation.

$$P(\mu_i \mid \mathcal{P}_j) = exp(\frac{\mid A_j - A_j^\star \mid^2}{\sigma_j}) \tag{2.4}$$

The matching rate of a model depends on each matching rate of the peaks. Even if these peaks have relations together as their summation generate the sounds, their independency will be taken as main hypothesis to avoid looking at the relation. At the end the probability for each model j to match on the measure will be computed by using the probabilities of the i peaks:

$$P(\mu_i) = \prod_{k=1}^{j} P(\mu_i \mid \mathcal{P}_k) \tag{2.5}$$

The final estimation of the road–tire friction coefficient is computed by using Markov chains. For the given starting configuration of probabilities, the final state is determined by the exploration of the convergence of the Markov matrix. Beforehand, the transition probabilities required for the formalism are determined linearly decreasing with the distance between the two μ classes. The next equation defines the final weights of the μ-classes by using the transition matrix for the six different classes and the identity matrix I_6 of size 6.

$$\begin{bmatrix} p_1 \\ p_2 \\ p_3 \\ p_4 \\ p_5 \\ p_6 \end{bmatrix} = \lim_{x\to\infty} \begin{bmatrix} P(\mu_1) & \frac{5(1-P(\mu_1))}{15} & \frac{4(1-P(\mu_1))}{15} & \frac{3(1-P(\mu_1))}{15} & \frac{2(1-P(\mu_1))}{15} & \frac{1(1-P(\mu_1))}{15} \\ \frac{5(1-P(\mu_2))}{19} & P(\mu_2) & \frac{5(1-P(\mu_2))}{19} & \frac{4(1-P(\mu_2))}{19} & \frac{3(1-P(\mu_2))}{19} & \frac{2(1-P(\mu_2))}{19} \\ \frac{4(1-P(\mu_3))}{21} & \frac{5(1-P(\mu_3))}{21} & P(\mu_3) & \frac{5(1-P(\mu_3))}{21} & \frac{4(1-P(\mu_3))}{21} & \frac{3(1-P(\mu_3))}{21} \\ \frac{3(1-P(\mu_4))}{21} & \frac{4(1-P(\mu_4))}{21} & \frac{5(1-P(\mu_4))}{21} & P(\mu_4) & \frac{5(1-P(\mu_4))}{21} & \frac{4(1-P(\mu_4))}{21} \\ \frac{2(1-P(\mu_5))}{19} & \frac{3(1-P(\mu_5))}{19} & \frac{4(1-P(\mu_5))}{19} & \frac{5(1-P(\mu_5))}{19} & P(\mu_5) & \frac{5(1-P(\mu_5))}{19} \\ \frac{1(1-P(\mu_6))}{15} & \frac{2(1-P(\mu_6))}{15} & \frac{3(1-P(\mu_6))}{15} & \frac{4(1-P(\mu_6))}{15} & \frac{5(1-P(\mu_6))}{15} & P(\mu_6) \end{bmatrix}^x \times I_6 \tag{2.6}$$

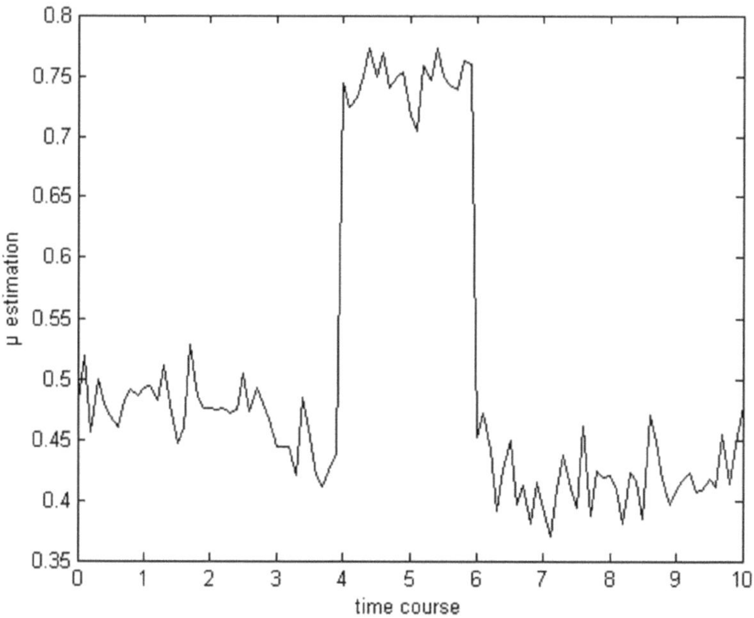

Fig. 2.20. Results of measure on cobblestones with a punctual patch

The example in figure 2.20 shows the right recognition of cobblestones with a punctual patch. Even if the patch is an abrupt change of the road surface, it will be detected only with a small latency time that corresponds to the time needed for the Fourier analysis to measure the new stabilized frequencies.

2.4 Local control of the predictive measures

In normal situation the checking algorithm is looking at the consistency of the results coming from both predictive estimation with a camera and reactive measurement with a microphone. The comparison of the results can be done when only the vehicle has been over the point of measure. As long as the measures are corresponding or the predictive data are a little bit lower than the reactive data, the predictive way of measurement gets the priority. Indeed if μ is underestimated, the vehicle safety will not be in danger as the vehicle dynamics will just be reduced.

But if the predictive measures overestimate μ, a dangerous conflict will appear. It may be possible that a small road portion has been wrongly analyzed when its surface is not wide enough. In such cases several types of textures have been matched together and the results tends to be estimated as a normal macadamized road ($\mu \approx 0.65$) as it can put emphasis on some

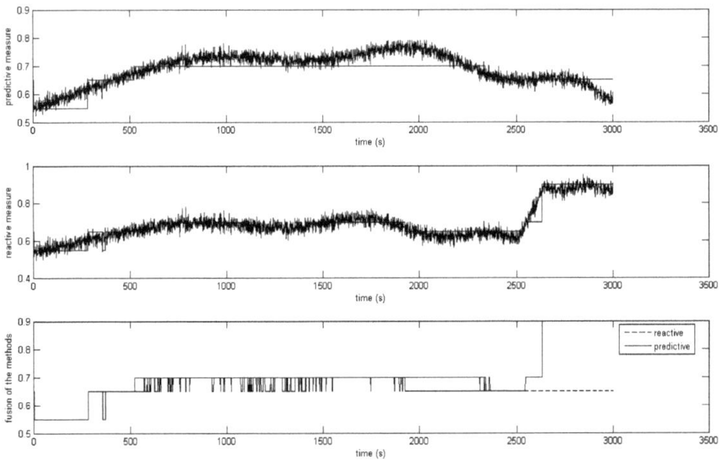

Fig. 2.21. Fusion of the predictive and reactive μ-estimation

local textures. In such cases, temporal errors have no real impact on the vehicle safety as their ranges are not wide enough to destabilize the vehicle, for example. Therefore a counter is integrated into the checking algorithm and it will switch off the predictive measurement only after a distance of 20 m.

On the next example (figure 2.21), predictive and reactive measures are presented for a daily test on a humid speedway with bad macadamized surface (μ is here about 0.7), as estimation and as class. On the last sub-figure the difference between reactive and predictive measures can be detected after the time 25 s. Actually the predictive measure has not detected the surface change and will not provide a correct class.

The example depicted in figure 2.21 shows also the limits of the estimation of the μ-coefficient. Actually, one can see the quick and continuous changes of the μ class by adding some noise to the estimations. Thus, the virtual driver is supposed to upgrade the minimal distances to other vehicles, the safe speed in a curve etc. depending on this noisy μ-class. Consequently, the assistant systems based on this data will have the risk to give highly changing output. All in all, almost each function in the vehicle will have after-effects because of the μ-estimation instability.

2.5 Pros and cons of the estimation methodology

Even if filters are added to the different μ-estimation, or hysteresis are used for the μ-class changes, the final information can be used only under probabilistic condition. Indeed, plugging safety relevant functionality on top of such estimation is really critical. Using additional redundant estimations with other physical methodologies is fundamental to decrease the risks of false detection. For example, indirect measures with the time, weather information etc. can be used to verify the results, or a vehicle observer can be used by looking at the wheel states over the ABS sensors and the engine torque control. However, the number of estimations is important, it will not be possible to ensure a safety relevant quality as failures may still happen. Moreover a cost problem will occur in this case because redundant measurements will generate costs overhead. Hence, two philosophies will appear: one will try to have the best overlaps of sensors' ranges; the other will try to have the best set of dedicated sensors. Any way, the method based on microphone is stable enough to improve the stability control. Furthermore, such device will be cheap enough for mass production contrary to improved vehicle observer that requires high end platform.

3

Actuators and drive train architecture

Abstract. The amount of different actuator topologies and their characteristics make the design evaluation validation of an architecture and above all the calibration of a virtual driver very challenging. Thus the use of a central entity that integrates the torque management (positive and negative values) and synchronizes the actuators by themselves makes the executive level unique to the command level independent of the platform configuration. The premisses for this concept come from the integration of the ACC and the hybridization of the drive train. The first one has to manage both combustion engine and brakes, thus it has to get an access to a torque strategy, which was until now within the combustion engine control unit. The second one allows several actuators possibilities for a same request, which requires an actuator strategy.

Based on this centralized torque management, the platform reliability will be improved over software integration for the actuators control, and after that replication of the ECUs will cost savings. Finally actuators integration will provide new degrees of freedom.

3.1 Migration strategy to a full safe drive-by-wire platform

The drive-by-wire technology replaces the mechanically linked actuators with actuators controlled only electronically over the network. For example, steer-by-wire removes the steering column between the steering wheel (sensor) and the steering actuator. As there is no mechanical back-up anymore, redundancy of the sensors, networks, actuators and even energy supply is required to avoid loss of actuator control. Hence drive-by-wire is the integration of brake-by-wire, steer-by-wire, power-by-wire (or also called eGas) and requires an automatic transmission.

Upgrading to drive-by-wire technology facilitates the vehicle engineering and its reliability, even if the acceptance from the passengers may be negative at the beginning due to this loss of mechanical back-up.[1] First new ergonomics are made possible by removing the steering column. Moreover symmetry for the chassis platform is not required anymore for right-hand and left-hand vehicles as it is the case today, further lowering localization of engineering costs. Furthermore, the driver defines the target and the system executes the control. He/she does not need anymore to perform the vehicle control over the mechanical parts linked to the actuators (e.g., wind compensation). Indeed he/she may have only to set the maneuver to perform and a substituting co-pilot can realize it semi-autonomously.

In addition, the new high-accuracy actuators such as the wedge brake (EWB) improve latency time, state control for shorter stopping distance and stability control. Indeed reliable real-time vehicle model (as described in the next chapter) cannot be extracted correctly for hydraulic brakes and mechanical steering actuator, but with drive-by-wire the determinism of the actuators' behavior is drastically improved. Finally wires can be saved through the integration of smart actuators, which improves the overall vehicle reliability as connectors are one of the main sources of failures.

In practice, from this point on autonomous vehicle will be possible (for the drive train part) as no human may be required anymore to take over control of the mechanical back-up in case of single electrical failure.

The migration to the drive-by-wire technology can be performed over two mechanisms. The first possibility is to simply replace current actuators with drive-by-wire components. In this case, steer-by-wire would be first in the integration list due to ergonomic improvements. These substitutions on decentralized architecture cannot improve slightly the vehicle dynamics as the same control strategy has to be used. A second introduction scenario seems more realistic because it is based on the integration of longitudinal and lateral controllers within one central ECU. With this centralized architecture, synchronization problems can be reduced within a software (and no more over different ECUs and networks) improving the vehicle validation since this ECU works as gateway between command level (HMI, assistant systems) and execution level. That is why the migration strategy to drive-by-wire can only be performed by following the strategy described below and illustrated in figure 3.1:

1. Functionalities have to be integrated within servers connected to a backbone to limit the amount of ECUs to duplicate and to improve the reliability (with lower sources of failure): drive train controller (engine/hybrid control with stability control), virtual co-pilot (assistant systems), driver working place (safety-related HMI such as command grabbing over pedals, steering wheel, and feedback management, instrument cluster), passengers

[1] Do not forget that airplanes from companies like Airbus of Boeing are based on the same technology named fly-by-wire for a while.

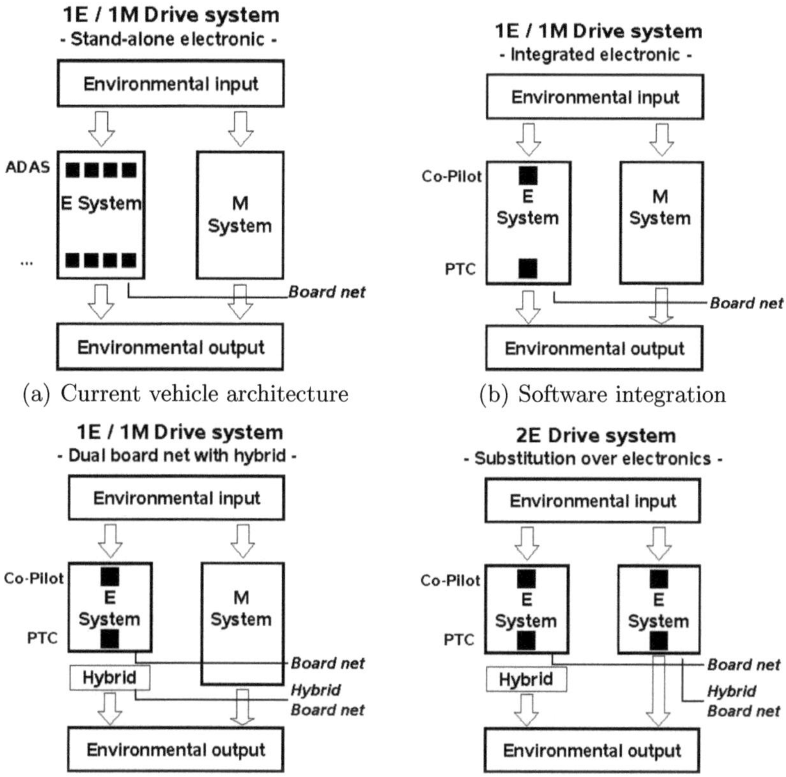

(a) Current vehicle architecture

(b) Software integration

(c) Dual board net with hybrid

(d) Electronics replication

Fig. 3.1. Migration strategy for the integration of a drive-by-wire platform

management (e.g., HVAC) and body control (e.g., high-voltage units, beams).

2. Integration of the hybrid drive train to get access to two different redundant energy supplies prior to the drive-by-wire introduction: 12 V for standard units and high voltage for special actuators such as first generation of steer-by-wire for trucks due to high torque requirements. At this step a parallel hybrid is sufficient as kinetic output is required.
3. Replication of the ECUs to replace the mechanical back-up with a low-cost electrical lane. Since the hybrid drive train has already been introduced, there is no need for additional energy supply. However, the actuators have only been replaced by brake-by-wire (the EWB requires 12 V thanks to the wedge concept) and steer-by-wire (high torque generation required here with at least 36 V with peaks up to 100 A) within the architecture.
4. The integration of actuators within the wheels can be prepared by using temporarily a torque vectoring. By introducing an additional clutch to

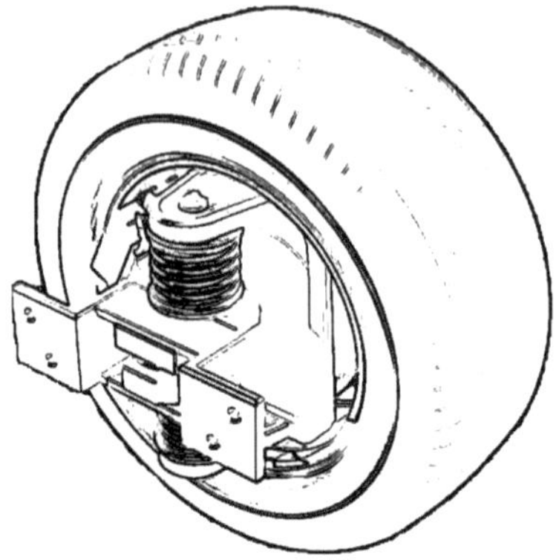

Fig. 3.2. In-wheel integration of the actuators

manage the torque ratio for front/rear axle and a differential for the front axle, it will be possible to bring independently the optimal torque to each wheel.

5. The integration of actuators within the wheels can be performed now with one fail-silent braking actuator, one steering motor (split of the two redundant motors from steer-by-wire) damping actuator and torque supply within one eCorner (figure 3.2) with local intelligence within the upgraded wedgebrake unit as wheel unit. The torque vectoring is now replaced by a direct energy supply to the wheel. Hence it is possible to get new degrees of freedom for the chassis control. Nevertheless full hybrid is mandatory here as only electrical power can be supplied to the eCorner.

3.2 Drive train architecture

A centralized drive train architecture can be modeled by four layers as shown in figure 3.3. The first layer integrates the central controller itself. It becomes a vehicle command to perform from the command level (driver or assistant systems) and it determines what has to be performed by each wheel by means of longitudinal, lateral slips and vertical dynamics (based on the sky-hook principle). In parallel this central controller sends back the vehicle state and its dynamic reserve to the command level to let them optimally overcome its instantaneous dynamic limitations. To determine the wheel unit commands, the vehicle stability control will be computed first. Its outputs define the requirement of relevant correction by means of braking or damping variation and it will also realize a threshold of the propulsive torque request (e.g., traction control). Hence it can also set the energy requested for the whole vehicle for realization by the energy management. This function integrates the hybrid control to select the optimal way to use the combustion engine, the electrical

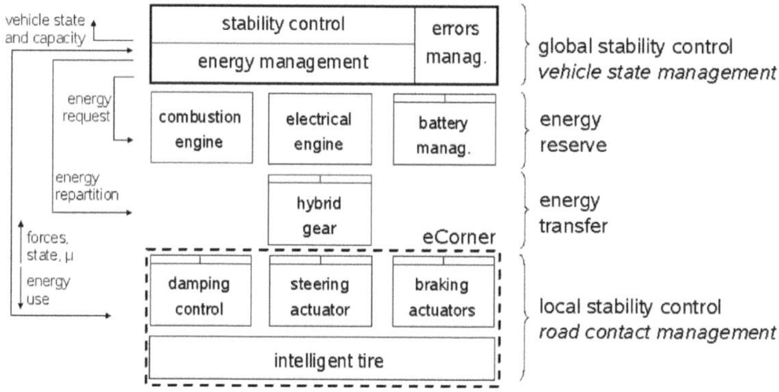

Fig. 3.3. Centralized drive train architecture with hybrid technology

motor and the brakes taking into account multi-objective criteria (level of acceleration or braking), the battery state and the environment (slope, etc.).

Once the energy management has defined the production command requests, they will be distributed to the different sources of energy such as the combustion engine, electrical motor or the battery and the alternator. The mechanic energy provided by these sources will be first added through a planetary gear and then the resulting torque will be distributed to each wheel unit depending on the individual needs (set by the stability control). This third layer corresponds to the energy transfer.

The last layer involves only the wheel units. For each unit, a local control unit integrating the slip control will optimally implement this command by combining propulsing torque from the energy sources, the resistive torque from the brakes and a control of the longitudinal and lateral slips with the steering actuator. Here the intelligent tire can send relevant information such as the road–tire friction coefficient or the tire pressure, which can be used locally for the damping or adapt the slips control directly.

In the end, this last layer named local stability control would be integrated within an eCorner. Hence only the global stability control would remain as central control for the strategy, the energy reserve, the energy transfer over dedicated wires and four eCorner modules.

3.3 Electrical integration with mechanical back-up

The first step of the migration strategy leads to separate sensors (pedals, steering wheel) to actuators while keeping a mechanical back-up in case of failure. Hence, the new technology introduced here only needs to be fail-silent. Let us take the current ESC as example to describe this concept. Within this ECU, both hardware and software are redundant within two parallel lanes. As soon as their outputs differ, the functionality is turned off and normal braking is performed without assistance with a minimum of 0.2 g as legally required. For the brakes upgrade, a power booster can be integrated on top of the hydraulic devices in order to get two inputs: a first input coming from the network with a safety check to perform external request such as those of the central controller, and a back-up input giving access to the brake pedal through direct wires. Hence a vehicle command can come and be performed by the brakes but in case of any failure, the driver still keeps a direct control.

The modifications are more likely to occur on the steering actuator since first versions of power boost are already available on the road. Up to now the only possibility to add an external steering correction available is active front steering (AFS) [71]. By integrating a planetary gear on the steering column, it integrates this request over an additional motor. However, some downside effects remain from the driver perspective. Actually, during a steering intervention, the corrective torque can only be put on the steering rack when the driver blocks the steering wheel correctly. Otherwise the rack will not rotate due to inertia, but the steering wheel with the driver's hand may rotate.

A correct solution needs to integrate a clutch on the steering column to separate the steering wheel to the rack, as in Fig. 3.4. Hence a direct steering actuator can perform the steering request issued by the central controller, and a feedback motor can emulate the vehicle dynamics over the steering column. Furthermore it can add haptic information in case of danger. As soon as a failure occurs, the actively opened clutch will close; hence a fail-silent system is integrated here too. The single drawback relates to the clutch closing phase. Indeed a static synchronization of the wheels and the steering wheel cannot be ensured. Thus the natural neutral position of the steering wheel cannot be guaranteed.

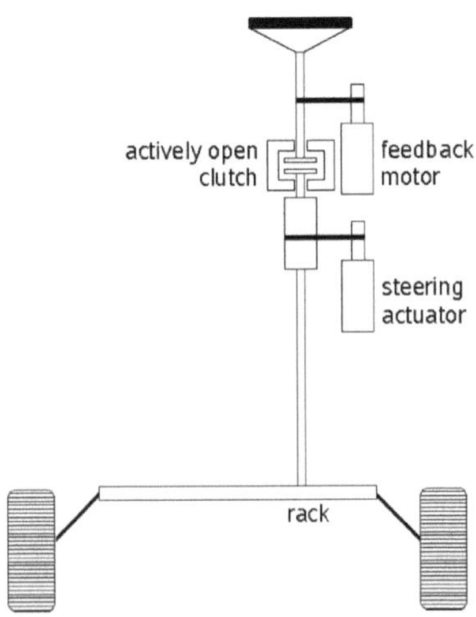

Fig. 3.4. First generation of steer-by-wire with mechanical back-up

3.4 Electrical replication

As described before, five main servers are required on a backbone. In practice these redundant ECUs will be plugged on the physical network as shown in Fig. 3.5. For reliability reasons [34], each ECU is fail-safe with single redundancy as in [17] and it is integrated twice to have a duo-duplex architecture as in [117].

Globally seen, the vertical symmetry corresponds to the redundancy requirements. As explained earlier, sensors, actuators, networks and even power supplies have to be integrated redundantly. Hence two quasi-identical networks are linking sensors (left side) to actuators (right side). Within such vehicle platform, CANbus cannot be used since it is not time deterministic (except with an additional TT-CAN layer), which leads to the use of FlexRay [109]. Hence it is possible to ensure the transfer of data without collision thanks to time slot allocations. Moreover FlexRay allows a higher bus load (up to 10 Mbit/s) which is mandatory for integrated chassis control or intelligent assistant systems.

Furthermore the layering concept presented in Chapter I.2 (figure 2.3) implies first a direct link between driver commands and actuator and after that the plug of additional levels, which generates first the drive train controller for the reactive active safety and then the assistant systems for the predictive active safety.

The first server (HMI server) is dedicated to the driver working place with the command sensors (HMI) mandatory on the redundant network. Two fail-safe sensors will be integrated here, which in turn leads to restrict this server to driver monitoring and the feedback management. These two functionalities are not mandatory, thus, it does not have to be fail-tolerant.

The second server (safety control) integrates together the assistant systems as virtual co-pilot and the passive safety (e.g., airbags). Since this server is safety relevant, it will be integrated redundantly to be fail-tolerant. Furthermore this server is connected to a dedicated FlexRay network for smart

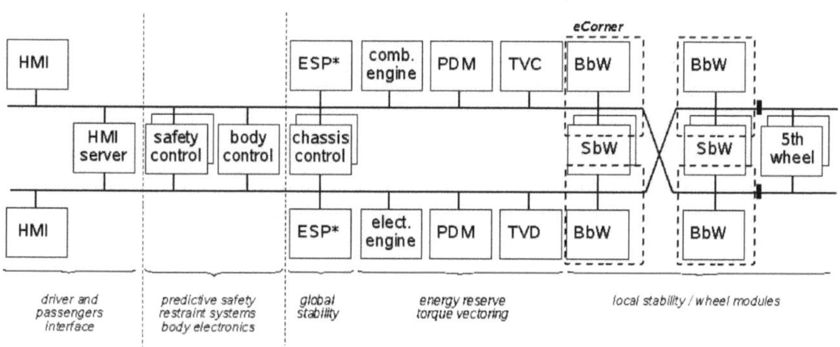

Fig. 3.5. Physical integration of the drive train

external sensors such as radars, navigation system or cameras with image processing platform. Transmitting a high data volume for information related to the surrounding environment, the network load has reached about 30%. The third server is the body controller as usual.

The last server is the drive train controller used as global chassis control. Behind this server, two yaw rate sensors (ESP*) are used. The combustion engine is only linked to one FlexRay network as it is not a mandatory actuator; but the power distribution management (PDM) which includes one battery each time has to be redundant. Then the actuators such as the brake-by-wire and steer-by-wire are cross-connected. The links over diagonal are chosen for stability purpose in case of loss of connection: in any case there will be one wheel per axle and one per side, which makes the vehicle stable. Last but not least, connections to the trailer are still required as a fifth wheel.

The next steps of the migration strategy require a hybrid power train. The electrical engine will be linked to the network that has no access to the combustion engine in order to have one engine per network in case of loss of network. Then the torque vectoring with clutches (VTC) and differentials (VTD) will be integrated, each one on a different network. The choice of the differential on the network with the combustion engine depends on whether the vehicle is rear or front wheels driven (once again just in case of loss of network to have a back-up with former topology).

For the validation of the vehicle architecture on open road, it is mandatory to get the vehicle stopped at least after the second failure. A reconfiguration algorithm has been introduced here to get the high-level servers such as the safety control re-loaded as chassis control. Thus up to eight failures can be tolerated until there is no more ECU to reconfigure as chassis control.

In any case, the braking actuators have to be synchronized. When the first diagonal loses its network connection, its brakes have to be opened in order not to disturb the vehicle stability with their full braking. After that, the other diagonal has to know that its braking actuators will have to perform a full braking when they lose their connection too.

An equivalent problem happens for steer-by-wire, where two motors (one per FlexRay network) are synchronized. When one motor loses its network connection, it has to be passive in order not to disturb the other motor. Furthermore the motors are developed powerful enough to override a bad passivated motor anyway. However, the passivation of the second motor has to ensure the vehicle staying safely on the road while stopping, which may not be the current steering angle (e.g., during a lane changing). Thus, the concept of command window (see Chapter II.5) will be introduced here to define this safe range.

4

Vehicle dynamics model

Abstract. The vehicle dynamics model is based on models of the actuators and the dynamic model of the vehicle. Through online adaptation of the actuators maximal dynamics, their transfer functions will describe their behavior and especially their physical limits. After the road–tire friction coefficient μ has been identified, the other relevant parameters are the energy state and the position of the center of gravity.

4.1 Modeling of the actuators

There are three kinds of actuators to integrate in this model. After the actuators with continuous state considering the requested approximation like the combustion engine, braking units or retarder, the gear and clutch aggregate will be modeled with its discontinuities. And finally a pure geometrical model will be used for the differential.

4.1.1 Modeling the dynamics of a unit with a second-order transfer function

Generally the generic models of transfer function have to use the lowest amount of parameters to adapt to the complexity detail of the modeling to minimum number. Moreover due to difficulty in identification of robustness, a too important level of details will not bring relevant information. For the actuators with a continuous dynamics, three physical parameters are relevant: latency time, rate, and maximal capacity. Hence these actuators can be described at least with a second-order transfer function as shown in figure 4.1 for the maximal engine torque. In this figure, the engine torque can propel the vehicle with a good μ (the case labeled one). But for the case labeled two, a lower maximal force can be transferred from the vehicle to the road because of the limitation of the friction coefficient. Thus the maximal acceleration will ever be lower for the same actuators.

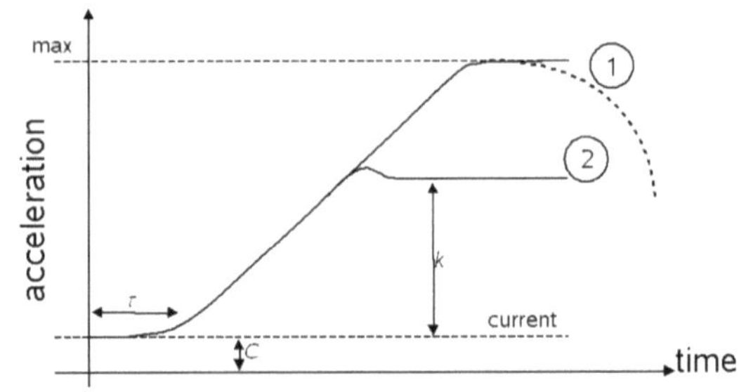

Fig. 4.1. Possible response for a full acceleration

Equation (4.1) defines in Laplace space the form of a second-order transfer function $\mathcal{F}$ with c the current state, k the gain, ξ the damping, T the pulsation, and τ a delay. Normally τ is not required but it will be used here anyway to modify the delays without having to compute the full transfer function again.

$$\mathcal{F}(p) = c + (\frac{k}{1 + \frac{2 \cdot \xi}{T} p + \frac{1}{T^2} p^2}) e^{-\tau \cdot p} \tag{4.1}$$

4.1.2 Non-iterative identification of the dynamics of units with continuous state

The states of the engine, the steering unit and the brakes are always stabilizing without important overshoot. The engine torque has a maximum level which can be reached for an optimal couple engine rotation and air–fuel ratio for gasoline and turbo for diesel. This optimum can be seen in figure 4.2(a) that shows an example of look-up table for an engine. For the steering unit, the maximal steering angle will be set as the physical one. Actually the lateral sliding has normally to be taken into account.

The computation steps are performed as following:

- Computing the gain k as the minimum after the rise.
- Identifying the delay as the time until amplitude is varying.
- Finding ξ and T.

For example, the method of Broida is adapted to make directly this identification without overshoot. It approximates a second-order transfer function as a first-order transfer function with a delay. The delay given in exponential form will be approximated by a Taylor series near zero (equation 4.2) when the delay is small. In normal cases such delay is roughly 0.3 s, which is small enough to validate this approximation.

$$F(p) = \frac{k}{1 + \tau_1 \cdot p} e^{-\tau_2 \cdot p} \approx \frac{k}{(1 + \tau_1 \cdot p)(1 + \tau_2 \cdot p)} \tag{4.2}$$

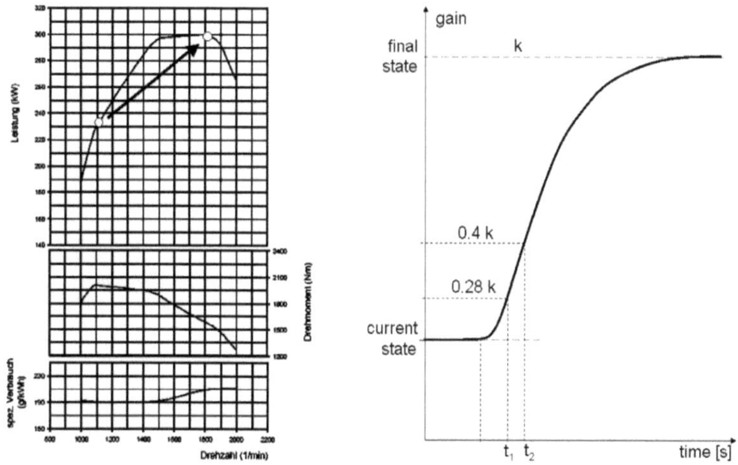

(a) Engine look-up table for an engine OM 501 LA from Mercedes-Benz

(b) Position of the characteristic data for the identification with the Broida method

Fig. 4.2. Look-up tables for the determination of the maximal dynamics

The identification requires characteristics similar to those shown in figure 4.2(b). The general curve uses two specific times and values of the Broida method, t_1 and t_2 :

- t_1 is the time when 28% of the rise is reached.
- t_2 is the time when 40% of the rise is reached.

Then τ_1 and τ_2 are identified:

$$\begin{cases} \tau_1 = 5.5(t_2 - t_1) \\ \tau_2 = 2.8 \cdot t_1 - 1.8 \cdot t_2 \end{cases}$$

Finally the physical parameters T and ξ are calculated. The time before stabilization at 95% reached can also be computed directly as the ratio $3/(T \cdot \xi)$.

$$\begin{cases} T = \sqrt{\frac{1}{\tau_1 \times \tau_2}} \\ \xi = \frac{T(\tau_1 + \tau_2)}{2} \end{cases}$$

This method can be applied directly for the braking units and the steering unit as their maximal capacities are defined. Figure 4.3 shows an example of braking response for a wedge brake with the identification of the transfer function and the simple model with only a rate limitation.

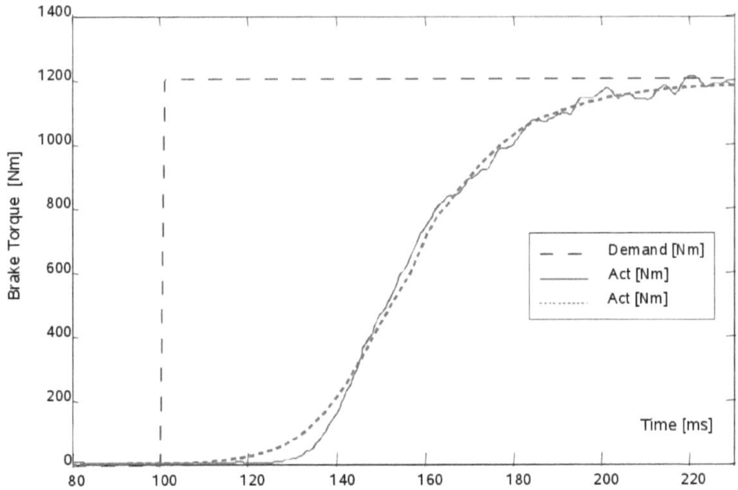

Fig. 4.3. Typical response of a brake-by-wire actuator

4.1.3 Identification of the dynamics of the retarder

The retarder can be integrated into heavy goods vehicle to filter the dynamics of the engine control. With this system, the controller of the butterfly valves will be filtered to avoid too high acceleration and high fuel consumption. The modeling is based first on a look-up table defining the maximal torque range depending on the butterfly valve angle and after that on a second-order transfer function which is statically determined.

4.1.4 Iterative identification of the gear and clutch dynamical model

Time model

An automatic transmission is mandatory since the synchronization of the different actuators is done by the central controller of the executive platform: if otherwise a human action would have to be requested for a gear shift. If the clutch is locked, the transfer function of the gear and clutch is based on the differential equation of the friction dynamics of clutch disks. The dynamical loss of torque due to the disks will be neglected as they become to be important only below 90 km/h where the clutch in a passenger car is always unlocked and stays artificially unstably opened for faster response. Hence only the extreme static gain of the gear is used here by neglecting the drive-shaft torsion. During a gear shift shown in figure 4.4(a) for the shift from the gear 7–9, a time model exists depending on the gains and the theoretical opening time of the gear.

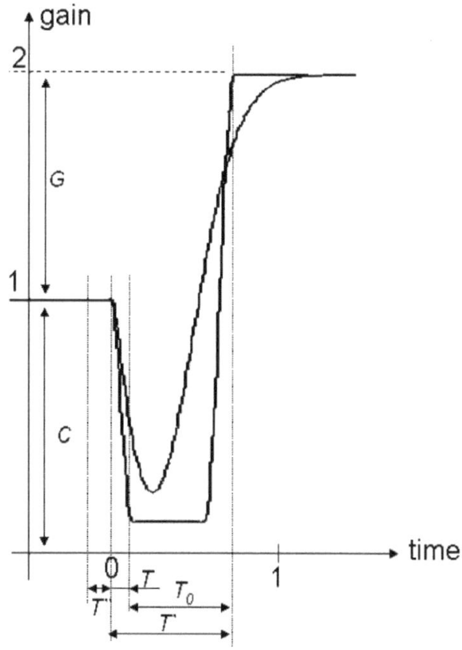

(a) Identification of gear and clutch transfer function

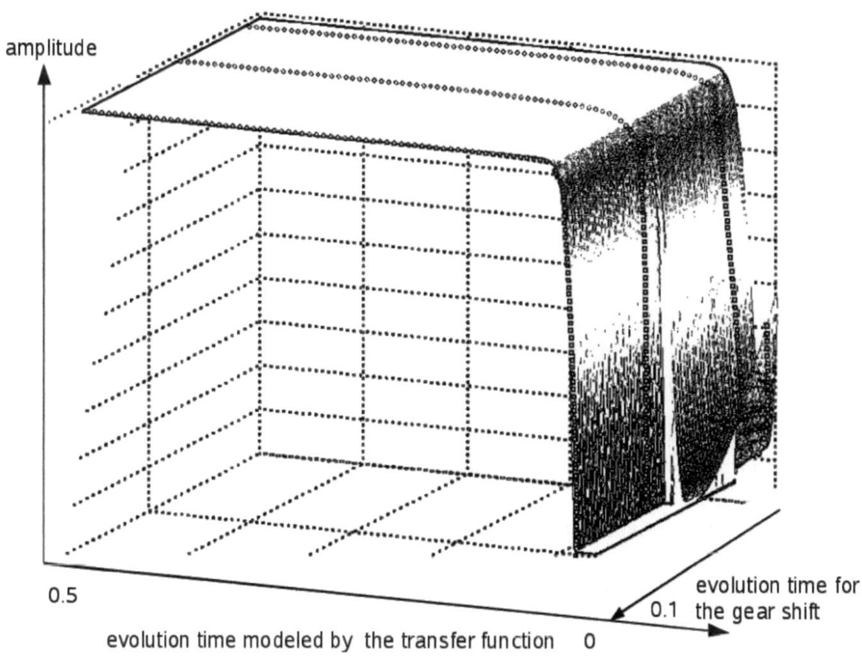

(b) Evolution of the gear shift temporal model during a shift

Fig. 4.4. Models used for a single gear shift

After that, this static pattern will be translated over the time to give the actual temporal model. In this case the gear and clutch unit can be modeled by three consecutive parts.

1. Opening the clutch with gear shift and closing the clutch. The specific identification of the transfer function will be explained in the next paragraph.
2. Gear shift and closing the clutch. During the gear shift, there is no propulsion torque and it implies a horizontal tangent of the torque gain as this corresponds to the local minimum of torque gain. Therefore the identification is also based on the Broida method explained in the section 4.1.2.
3. Clutch closing. A first-order transfer function can be used because there is no horizontal tangent at the beginning. The method for the identification will not be described here, but is based on the same principle as explained in [116]: a tangent is used directly to define the dynamics.

The switch from one transfer function model to the next one implies discontinuities as shown in figure 4.4(b), where abrupt changes can be seen on the right side (corresponding to the time where the gain is almost zero). For example, when the clutch is opened, a second-order transfer function is used to model the latency time where the clutch stays open until it closes. But as soon as it starts closing, its model will be a first-order transfer function. Actually the first-order transfer function has a quasi-vertical tangent at the beginning, which is not compatible with the former dynamics model. These discontinuities can be attenuated but it implies the use of an iterative identification method that needs many control points for the matching. This iterative identification will not be done here to avoid long computations and risk of stack overflow within the embedded platform because of uncontrolled amount of iterations.

One gear shift model

The gear and clutch temporal model is corresponding to two successive Heaviside functions (H)[1] with delays. The data necessary to identify these functions are described in figure 4.4(a):

- Current state C: the depth decreasing from current state to zero value.
- Gain G: the difference between the current gain and the gain of the next gear.
- First delay time value T'': for the first Heaviside function before the clutch starts the opening.
- Second delay time value T: for the time to get the clutch opened.
- Last delay time value T': for the time where the clutch is opened.

[1] A Heaviside function is corresponding to zero before the activation and a constant after the activation.

The response of the system with a Heaviside function is

$$
\begin{aligned}
\mathcal{L}(H(t-T-T'')+H(t-T'-T'')) &= (\frac{-C}{p}e^{-T\cdot p}+\frac{(C+G)}{p}e^{-T'\cdot p})e^{-T''\cdot p} \\
&= \frac{e^{-T''\cdot p}}{p}(-C\cdot e^{-T\cdot p} \\
&\quad +(C+G)e^{-T_0\cdot p})e^{-T\cdot p}) \\
&= \frac{e^{-T''\cdot p}}{p}TF(p)
\end{aligned}
$$

Where $T_0 = T' - T$ and with

$$TF(p) = -C\cdot e^{-T\cdot p} + (C+G)e^{-T_0\cdot p})e^{-T\cdot p} \tag{4.3}$$

Then the transfer function is transformed by approximating the exponential with a Taylor development on 0 relative to the delay to get a transfer function in standard form.

$$e^{-Tp} \approx \frac{1}{(1+\tau)^{\frac{T}{\tau}}}$$

This method gives results with 15% accuracy if τ is small (generally $0.100s$). But the form will also be very complex if a high precision of the matching on the Heaviside functions is needed: the size of the numerator and the denominator can be very important, about 10–30 elements to get accuracy better than 90%.

Unfortunately, the time of an automatic gear shift is not always constant. In practice, depending on the difference in speed of the two clutches and the engine torque, a slipping behavior can be observed time to time, which will prevent a good grip. In such cases, the algorithm for automatic gear shift will re-open the clutch and try later. Therefore a probabilistic approach for the transfer function of the gear and clutch will be necessary to get the probability of the available torque at each time depending on the probability that the clutch will be close. The functions are continuous in the time space but not in the parameters space of the differential equation (Fig. 4.5).

Sequential gear shifts model

If the vehicle requires a higher torque than possible or to generate as a engine brake, it will be possible to have several sequential gear shifts. In these cases the method presented previously can be reused and extended for the (i) other next shifts with a delay.

$$TF = \sum_{i}^{n} TF_i(p) \times e^{-T_i\cdot p}$$

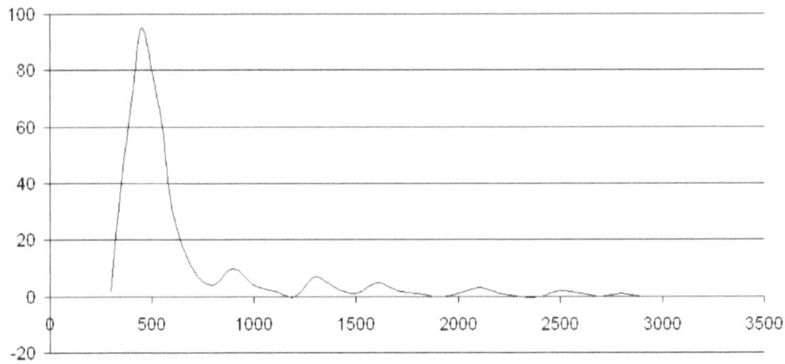

Fig. 4.5. Statistical distribution of the gear shift time for an Actros at 50 km/h

An accurate approximation of the discrete states of the gear box is difficult to obtain with second-order transfer functions as in figure 4.6. Indeed a problematic compromise has to be done between convergence speed and interdiction of any over-shoot. There exist different alternatives described in the following. For example, a successive use of first-order transfer function can model the abrupt changes with a better accuracy or it is also possible to define directly the model as a list of constant states with the couple

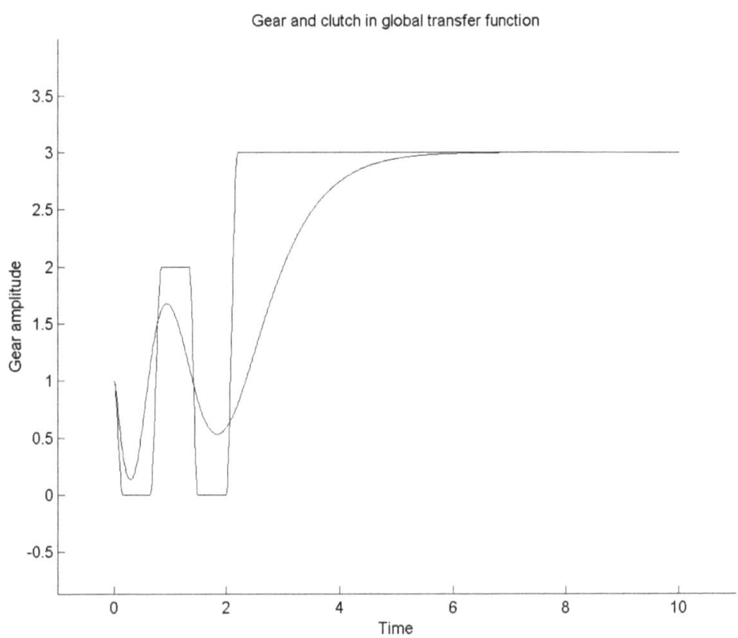

Fig. 4.6. Identification of a gear shifts sequence

ratio-operation time. However, these alternatives can not be translated after in terms of global transfer function as used previously. That is why they were not integrated here, because model cohesion is much more valuable than accuracy of sequential gear shifts model (low recurrence). Thus no speed change is used for the short-time control strategy for the command level.

4.1.5 Non-iterative identification of the differential model

The task of the differential is to distribute the torque on the right and left propulsive wheels depending on the road curvature. If not some damages can occur on the axle, due to the difference in wheel speed. As the curve radius is not internally defined the steering angle will be used to approximate it

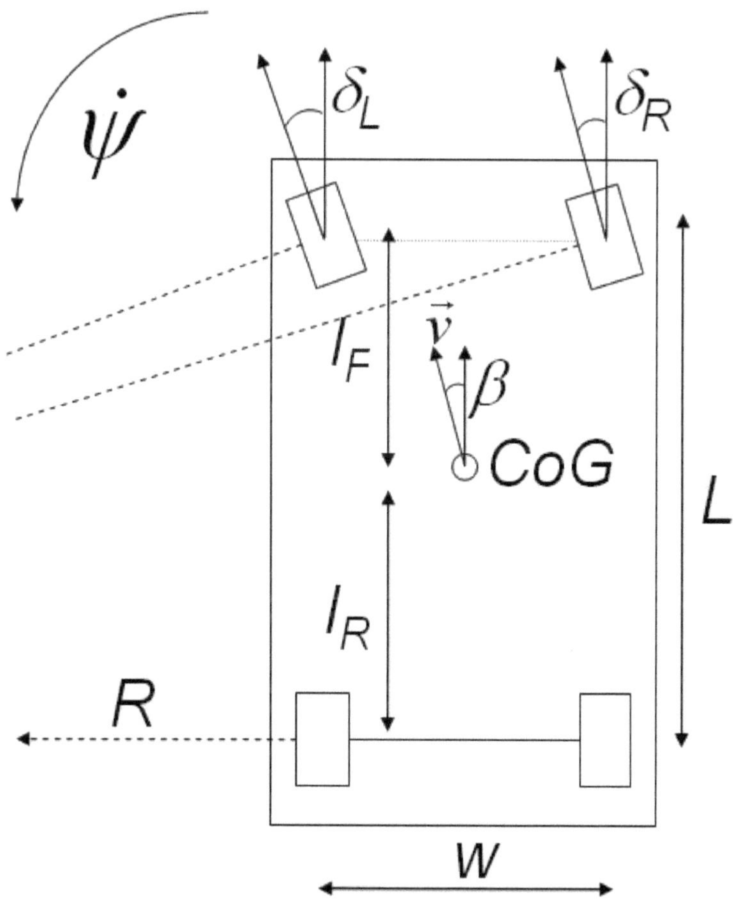

Fig. 4.7. Ackerman model for a vehicle

by using an Ackerman model in figure 4.7. The steering angle δ is defined as $atan(L \;/\; (R+w/2))$ when neglecting the lateral acceleration influencing the under-steering or over-steering behavior of the vehicle.

First, let us define the left wheel r_L with a speed of v_L and the right wheel r_R with a speed of r_R. The calculation of the torque repartition ($\rho_{R,L}$) on the wheels is given as follow, depending on the instantaneous path curvature R for the left wheels:

$$\rho_R = \frac{v_R}{v_R + v_L} = \frac{\frac{(R-\frac{l}{2})\delta}{t}}{\frac{(R-\frac{l}{2})\delta}{t} + \frac{(R+\frac{l}{2})\delta}{t}} = \frac{R-\frac{l}{2}}{2R}$$

And $L = R \times \delta$, with $\delta = \frac{\delta_L + \delta_R}{2}$:

$$\begin{cases} r_R = \frac{1}{2} - \frac{l\cdot\delta}{4\cdot L} \\ r_L = \frac{1}{2} + \frac{l\cdot\delta}{4\cdot L} \end{cases}$$

By integrating the parameters of the transfer function for the steering angle (c_δ, ξ_δ, T_δ, τ_δ), the final transfer function is obtained:

$$\begin{cases} r_R = (\frac{1}{2} - \frac{l}{4\cdot L}c_\delta) + \frac{\frac{1}{2}-\frac{l}{4\cdot L}k_\delta}{1+\frac{2\cdot\xi_\delta}{T_\delta}p+T_\delta^{-2}\cdot p^2}e^{-T_\delta\cdot p} \\ r_L = (\frac{1}{2} + \frac{l}{4\cdot L}c_\delta) + \frac{\frac{1}{2}+\frac{l}{4\cdot L}k_\delta}{1+\frac{2\cdot\xi_\delta}{T_\delta}p+T_\delta^{-2}\cdot p^2}e^{-T_\delta\cdot p} \end{cases} \tag{4.4}$$

However, the computation of the torque from the braking pads to disengage the differential lock will not be performed here, as the modeling is too complex for any embedded integration.

4.2 Limitation due to electrical power

The vehicles are integrating more and more electronic control on the smart actuators as brakes and steering units especially with the X-by-Wire technologies. On one hand the accuracy is improved but on the other hand these units need a safe energy management delivering enough energy to each unit at any time. Otherwise the vehicle will not be able to steer or to brake.

Unfortunately the level of energy is limited by the capacity of the battery on board and the alternator capacity. Additionally this level depends on external factors such as the slope of the road, and the number of the electronic system switch on at the moment (energy strategy).

Therefore a new functionality has to be integrated into the vehicle to manage the distribution of energy. Moreover the limitation of the energy for the units will be considered at the command level to avoid any situation involving a need of energy higher than the vehicle capacity.

4.2.1 Model of maximal available energy

The state of the battery and the energy available from the alternator are measured over the time. The instantaneous lack of energy from the alternator (problem of alternator control) can be compensated during a short time by the reserve in the dedicated battery. Therefore the different actuators have no influence on each other because each one is only allowed to use its own back-up battery. But the problem of maximal available energy is happening when all the batteries need to load or at least one battery has failed.

Two differential models of the dynamic of both batteries [24] and alternator [45] are set into a Kalman filter with infinite horizon in [14]. This wide horizon permits to extrapolate the maximal energy available over the next second/minute. The extrapolation permits to know how long the battery can compensate the lack of energy from the alternator before no more energy is available.

The energy available can be used by all the electronic units in the vehicle in the best case. The core architecture such as the ECUs cannot be shuted down. Therefore, there is always a minimal level of energy to save for this usage. The rest of the units such as the braking, steering or the radio etc. is separated into several classes of energy integrity. For example, steering and braking actuators are corresponding to the first category, which cannot be delayed or switched off. For the other categories, the requests are sorted depending on their integrity levels, and they may have to reduce their needs or to postpone them in order to avoid such problem (Fig. 4.8).

4.2.2 Optimizing the energy capacity

Each unit in the vehicle has a pattern of the energy needs depending on the situation: temperature, current state, actions to realize. When these units are activated by the driver, their patterns are loaded and summated together as

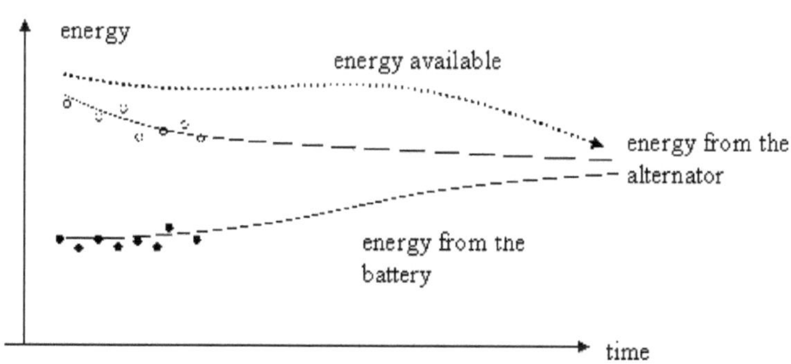

Fig. 4.8. Model of maximal available energy

in figure 4.9(a). All patterns have fixed amplitude in exception of the radio and air conditioning, which could adapt their level of energy since they are only integrated as comfort functions. Therefore the maximal energy request can be found. The algorithm checks that it will require more than the capacity available at this moment. Otherwise the vehicle will not be able to realize the driver commands.

In such case the algorithm still has some possibilities to achieve its tasks. If the maximal level of energy is due to a summation of different needs, it can try to postpone the actions of the non-safety relevant units as in the figure 4.9(b). The units are sorted by safety relevance to know which activity has the highest priority to be performed: brakes, steering unit, engine control, springs control, blinkers, light horn, horn, doors module, air cooling, radio etc.

4.2.3 Modifying the command to adapt it to the energy level

Right now in conventional vehicles, when too much energy is required, the saturation effect is managed on a reactive way since the system the command or the requested cannot be estimated in advance. Hence it is not possible to determine which elements will get troubles. If one wants to integrate a safe driver-by-wire technology, the energy request from the x-by-wire components

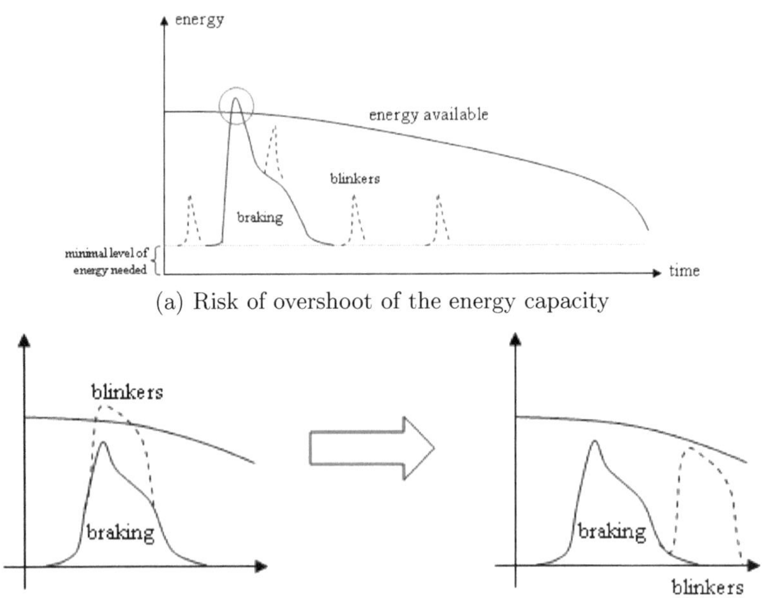

(b) Postponing energy needs to stays below the maximal energy capacity

Fig. 4.9. Use of battery model for the postponing of activities have to be ensured. Thus new methodologies for energy management as proposed here are trying to modify the energy requests when necessary to avoid the overall energy request reaching its limits. If one unit requires more energy than available, the energy management will not get the possibility to realize the command efficiently. Therefore the command will be adapted to realize the same effect at the end of the action but with a lower maximal energy need and a longer time of use.

The modification of the command will be done by modifying the size of the pattern. The models of the energy used by the engine, braking and steering units are defined as functions depending on the vehicle command example. The single parameter to work with is the level of the command. Therefore the value of the command can be modified to get the energy need equal to the maximal energy capacity. The modification of the command will also modify the duration of the use of energy. For example, by lower braking power, less energy is required and moreover the braking units need less time to open: they need energy during more time (Fig. 4.10).

4.2.4 Pre-compensation of the physical limitations

The modification of the command is not dangerous for the driver as long as the command keeps the system behavior within a comfort range. If it is not the case, it will reduce the performance of the vehicle (like longer distance to brake). In practice, this pre-compensation has to check that the vehicle can always brake at every moment without colliding with the vehicle in front. Therefore the maximal capacity of the energy management has to be taken into account in advance to avoid having the vehicle in such unsafe state. The safety limitations of the vehicle are only dealing with the engine, the braking, and the steering units. Thus for each predicted time the maximal capacities of the power pack unit, of the brakes and of the steering unit are computed separately. At the end, these three technical capacities limit the full capacities of the vehicle. Furthermore the system can determine all the possible commands or emulate them with a regression in terms of limited energy.

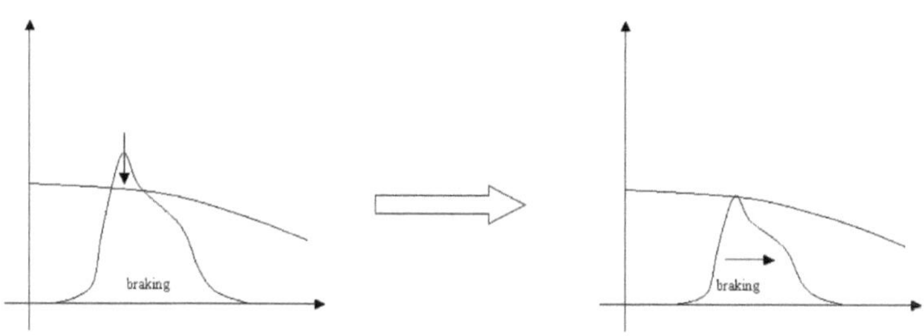

Fig. 4.10. Mitigation of the command to reach the maximal energy

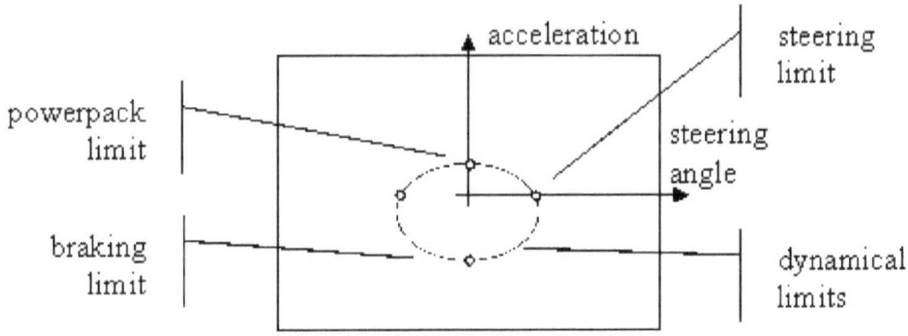

Fig. 4.11. Limits of the feasible commands depending on the maximal energy level

$$\begin{cases} E_{brakes} \approx U \cdot I_{brakes}^2 \\ E_{steering} \approx U \cdot I_{steering}^2 \end{cases}, I_{brakes} + I_{steering} = I_{max}$$

$$E_{brakes} = U \cdot I_{max}^2 - 2 \cdot U \cdot I_{max} \sqrt{\frac{E_{steering}}{U}} + E_{steering}$$

$$E_{brakes} = E_{max} - 2 \cdot I_{max} \cdot I_{steering} + E_{steering} \tag{4.5}$$

This regression defines the dependencies between the maximum capacities of the actuators. With this regression, it is possible to pre-compensate the level of braking or steering available depending on which battery is currently under load. Results are given in the figure 4.11 for a load of the batteries dedicated to the steering unit. Indeed, the vehicle cannot go out of this envelope. On top of that, this maximal energy level will decrease the force capacity that the actuators can generate. Thus this energy decrease will require modifications on the maximal state of the identification of the actuators in parallel to the modification on the maximal state due to the road tire coefficient.

4.3 Dynamics model

The disturbances due to the multi-body elements of the vehicle such as slosh motions into the tank or the cabin oscillations are not taken into account because only general information has to be provided.

The vehicle dynamics is determined through a bicycle model. Actually up to now this model is working as a differential equation to model the acceleration and the yaw rate of the vehicle. These differential equations can re-use the dynamics models presented previously to describe the extreme dynamics as another transfer function, which in turn will be sent to the command level to describe what the vehicle can realize.

4.3.1 Computation of the propulsive forces

On each wheel (i) with a mass $M_{wh.}$ the torques coming from the powertrain $\mathcal{T}_{pt}$, from the braking unit $\mathcal{T}_{br}$ and from the friction on the road $\mathcal{T}_{road}$ depending on the slope (α) and on μ give together the wheel speed given in equation 4.6. Furthermore, an additional torque coming from parallel hybrid can be added here. In a parallel hybrid, combustion engine and electrical engine can work in parallel and their torques are combined on a mechanical gear.

$$\frac{M_{wh.} \cdot r^2}{2}\ddot{\theta}_i(t) = \mathcal{T}_{pt} - \mathcal{T}_{br} - \frac{l}{2L}\mu \cdot \underbrace{M_{wh.} \cdot g \cdot cos(\alpha)}_{F_N} \tag{4.6}$$

This differential equation integrates the different transfer functions defined previously. The product of the equations (4.2), (4.3) and (4.4) defines the propeller torque $\mathcal{T}_{pt}$. If the propeller torque or the braking torque $\mathcal{T}_{br}$ is too high, a sliding effect of the wheel will happen. Thus the corrective function such as ABS and traction control will also have to be taken into account.

These torques on the wheel will generate a propulsive force ($\mathcal{F}_i$) acting on the vehicle depending on the normal force of the vehicle ($\mathcal{F}_N$) and on the capacity (k) to transfer this force that depends on the longitudinal slip (λ) and on μ as shown in figure 4.12. The standard generic curves defined by Pacejka will be used here without modifications.

$$\begin{cases} \lambda = 1 - \frac{\dot{\theta} \cdot r}{v} \\ \mathcal{F}_i = \mathcal{F}_N \cdot k(\lambda, \mu) \end{cases} \tag{4.7}$$

Unfortunately the integration of λ into the vehicle dynamics prevents any direct computation from the wheel torque to the vehicle speed because of their mutual dependency. Thus, the process of speed determination starts by changing the dynamics representation from Laplace space to temporal space.

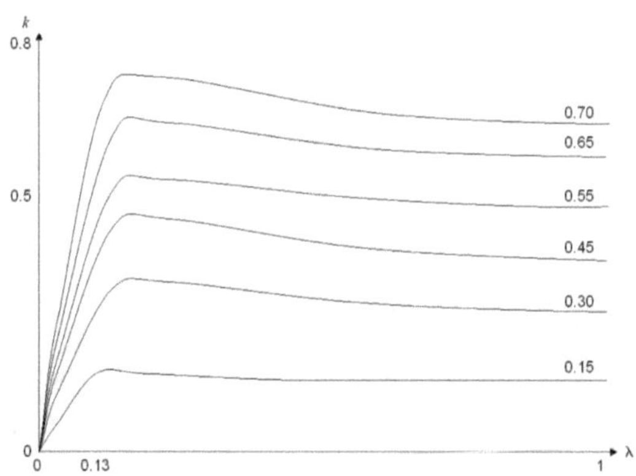

Fig. 4.12. Graph of k depending on (λ) and on (μ)

Then, the slip value is computed iteratively with the former speed value and the new wheel speed. At the end, the methods presented before are used once again to transfer the speed variation into Laplace space.

4.3.2 Computation of the vehicle dynamics

In this section, the aerodynamic forces (drag, positive, negative lift and ground effects) are neglected to simplify the computations. Their influences can be neglected anyway for standard passenger cars and heavy good vehicles at low speed. Furthermore the integration of the drag would generate a non-linear phenomenon that can be linearized for the current speed to be integrated within the transfer functions.

If the vehicle is only modeled as a single point, the vehicle acceleration γ can be extracted from the geometrical model presented in figure 4.13 with h the position of the center of gravity, $d_{front} + d_{rear}$ the wheel base and α the road slope.

$$\begin{cases} m \cdot \gamma = F_{x,front} + F_{x,rear} - m \cdot g \cdot sin(\alpha) \\ 0 = F_{z,front} + F_{z,rear} - m \cdot g \cdot cos(\alpha) \\ 0 = F_{z,front} \cdot d_{front} - F_{z,rear} \cdot d_{rear} + (F_{z,front} + F_{z,rear})h \end{cases} \tag{4.8}$$

Thus, in normal case, the transfer function of the acceleration / braking capacity of the vehicle is defined with the summation of the forces generated by the four wheels torques.

$$\begin{cases} M_{veh.} \cdot \gamma = \sum_i \mathcal{F}_i - M_{veh.} \cdot g \cdot sin(\alpha) \\ \sum_i \mathcal{F}_i = \mathcal{F}_{r,l} + \mathcal{F}_{r,r} + \mathcal{F}_{f,l} + \mathcal{F}_{f,r} \end{cases} \tag{4.9}$$

$$v = \frac{-M_{veh.} \cdot g \cdot sin(\alpha)}{m \cdot p} + \frac{1}{m \cdot p}(4 \times TF_{br} + 2 \times TF_{pt}) \tag{4.10}$$

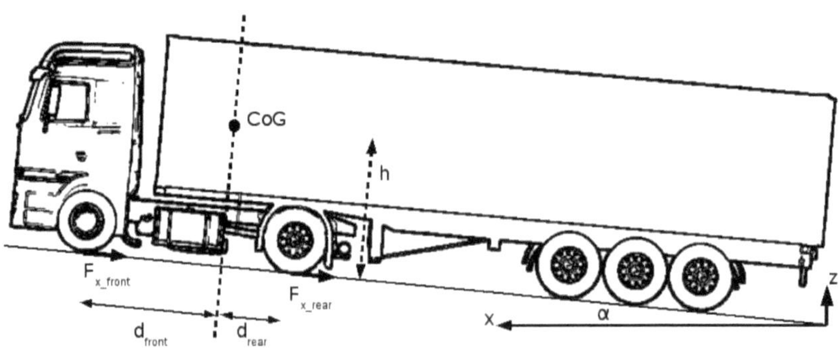

Fig. 4.13. Definition of the normal forces

On one side the acceleration will be defined as a second-order transfer function when the vehicle is purely braking. On the other side an accelerating vehicle with a gear shift will have its acceleration modeled as fourth-order transfer function, which is the product of second-order transfer functions for the engine and for the gear and clutch. Here the possibility to apply both braking and propelling torques at the same time will not be taken into account to avoid having a fusion of a second-order transfer function with a fourth-order transfer function. But this fourth-order transfer function can anyway be reduced with an iterative method to decrease its order down to 2 with loss of information, as in [91]. In this case, it should be possible to sum anyway the braking and propelling transfer functions.

But the maximal acceleration may be limited by the dynamical acceleration capacity of the vehicle. Depending on the position of the center of gravity and the vehicle mass, the maximal feasible acceleration can vary as it depends on the load forces on the wheel. Indeed, the dynamical axle loads, which is defined as followed and defined for the first time by Mitschke in [88], needs to be taken into consideration.

$$\begin{cases} F_{z,front} = m \cdot g \cdot cos(\alpha)(\frac{d_{rear}}{d_{front}+d_{rear}} - \frac{h}{d_{front}+d_{rear}}(\frac{\gamma}{g} + tan(\alpha))) \\ F_{z,rear} = m \cdot g \cdot cos(\alpha)(\frac{d_{front}}{d_{front}+d_{rear}} - \frac{h}{d_{front}+d_{rear}}(\frac{\gamma}{g} + tan(\alpha))) \end{cases} \quad (4.11)$$

For a rear wheel driven vehicle, the maximal acceleration is defined when the force is full on the front axle ($Fx_{front} = -\mu \cdot Fz, Fx_{rear} = 0$). Thus it can be extracted from the equations (4.8) and (4.11):

$$\gamma_{max} = -g\frac{\mu}{1 - \mu\frac{h}{d_{front}+d_{rear}}}\frac{d_{rear}}{d_{front} + d_{rear}} \quad (4.12)$$

This vehicle dynamics will be transferred to other elements within the architecture. This pure τ delay is modeled with a single $exp(-\tau)$ in the Laplace space. But the method used before can also be re-used once again to generate an approximate final transfer function.

4.4 Use of the dynamics model

This chapter presented a dynamics adaptation method for real-time purpose in order to get information about the acceleration, braking and steering capacities. The identification depends first on the energy level for the actuators, whose response has to be adapted. After that, the theoretical response has to be adapted to the situation depending on the current road–tire friction coefficient. This identification describes the vehicle response for the maximal demands as usual. Hence it should be possible to optimize the controllers, which are integrated within the executive level, for the realization of the vehicle command. Indeed the inversion of these transfer functions would help to get quasi-identity transfer functions for the overall couple controllers actuators.

However, this concept would only work if the generic response will be compliant for every command. Unfortunately for commands with lower intensity, the vehicle response in terms of acceleration does not imply a gear shift or use of the retarder, for example. Thus, as the vehicle is a multi-body system, whose elements get their own algorithms for the synchronization, the quality of the identification for the controllers is not so important. But the most important added value will influence the command level. Indeed the controllers such as speed or distance control will get actual data about the executive level, which prevents having to set fixed parameters within these controllers.

Anyway, the dynamical model cannot be sent as transfer function everywhere. From time to time, a complex transfer function will not be useful when geometrical computation are performed, using a different representation space of the model. Hence a last requirement for the executive level may be to able re-writing the dynamical model with other parameters approximating this model: e.g. latency time, variation rate (as a constant here) and maximal capacity.

5

Performing the vehicle command

Abstract. The execution level will receive from the command level a command in terms of acceleration and steering angle to realize. It will have to choose the right actuators to perform this command optimally. There is not a unique solution to perform this command. Thus, a command window will be requested to limit the range that can be used to perform the action. After that an analysis of the command variation will be done to choose the actuator and to control the action.

5.1 Command range

In addition to the vehicle command coming from the command level, the executive level will receive a command range, which defines the maximal and minimal acceleration and steering angle allowed, from the assistant system. In normal case, these limits corresponds to the comfort range, but in emergency case it will correspond to the boundaries of the safety envelope. The powertrain can use this range to realize the vehicle command optimally. This flexibility is not incompatible with the needs to perform the vehicle command correctly. Actually, the powertrain controller needs this range (especially for the longitudinal control) to check to enable a gear change strategy. To check the influence of the gear shift, it can use directly the gear shift transfer function. As long as the minimum of acceleration during the gear shift is higher than the given limit, the gear shift is allowed.

Figure 5.1 depicts the first gear shift for an Actros MP-II at about 35 km/h. While the clutch is open, no propelling force can compensate the rolling forces and the wind effect. Thus, the speed will decrease shortly till the clutch is closed and the vehicle can accelerate again. Here the jump of the signal speed is due to the temporal change of the sensor (ABS). Actually when the clutch is closed at low speed, the speed measure is performed at the output of the gear. However, when the clutch is open, the speed measure has to be performed anyway at the wheels directly.

Two different command ranges are shown in figure 5.2. In the case 5.2(a), there is nothing behind the vehicle, so full braking is allowed. But in the other case, no braking is allowed because of the vehicle behind (e.g., during an overtaking maneuver). The gear shift will be possible in the first case but the second case will be possible only on descent. Actually in this case, the vehicle can still accelerate during the gear shift because of its inertia.

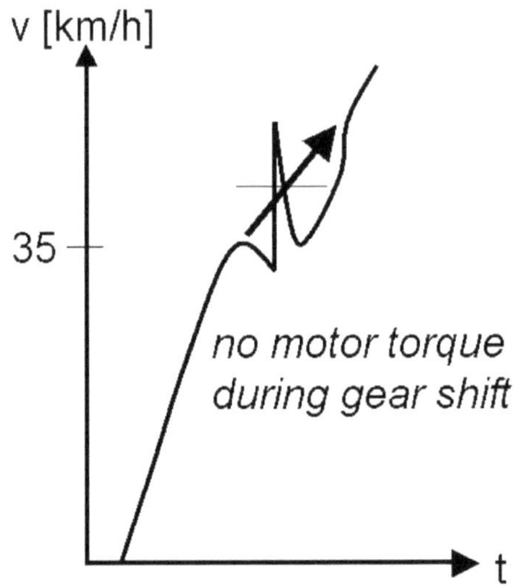

Fig. 5.1. Example of gear shift by an Actors MP-II from Mercedes-Benz

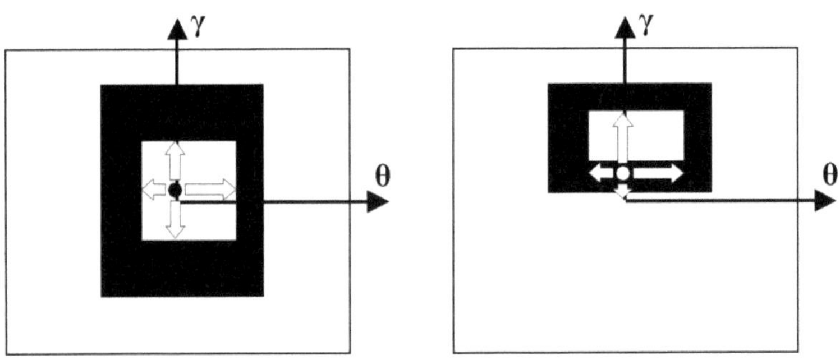

(a) Possible case of gear shift (b) Case preventing the gear shift

Fig. 5.2. Two examples where the minimal acceleration (γ) is relevant but not the steering angle (θ) for the gear shift control

5.2 Inverse computation of the actuators' command

As the engine control is slower than the brakes, its latency time will be used for the control as reference. It has been shown in the section II.4.3 how to get the transfer function of the powertrain, which describes the capacity of acceleration. This second-order transfer function can be approximated as a first-order transfer function with a delay τ, which is always relative small. [1] Furthermore, there is always a real solution because no overshoot happens. Furthermore, it is not allowed by the law to realize such overshoot since they can be dangerous. Thus, from this first-order differential equation, a control parameter α_γ can be extracted to define an acceleration control:

$$\begin{aligned} TF &= \frac{k}{1 + a \cdot p + b \cdot p^2} \\ &= \frac{k}{(1 + \alpha_\gamma \cdot p)(1 + \tau \cdot p)} \\ &\approx \frac{k}{1 + \alpha_\gamma \cdot p} e^{-\tau \cdot p} \end{aligned}$$

The controller will receive an acceleration request ω_γ and attempt to reach the corresponding vehicle acceleration (γ). The controller will adapt the resulting acceleration, which can be modified by a sensitivity parameter α_γ. It can be adapted depending on the vehicle mass, the acceptance on vehicle oscillations, etc.:

$$\frac{1}{\alpha_\gamma}\dot{\gamma} + \gamma = \omega_\gamma \tag{5.1}$$

Thus, the control force F_x to realize can be defined as following (here the resistive force will be added as $F_R(v, \alpha)$):

$$F_x = F_R(v, \alpha) + \underbrace{m \cdot \alpha_\gamma(\omega_\gamma - \dot{\gamma})}_{requestedforce} \tag{5.2}$$

The controller will either modify directly the engine throttle α_t (in practice the torque request) or it will apply a braking force. For simplification, a look-up table F_E defines the engine torque depending on the throttle and the engine speed, as in figure 4.2(a) in Chapter II.4.4. From this look-up table, the minimum graph will be reused to choose which actuator will be in request:

$$\begin{cases} Fx \geq F_E(\alpha_{t,min}, v) \longrightarrow f_E = F_X \\ Fx \leq F_E(\alpha_{t,min}, v) \longrightarrow f_B = -(F_X - F_E(\alpha_{t,min}, v)) \end{cases} \tag{5.3}$$

[1] Here the delay is always chosen as the smallest positive eigen value of the system.

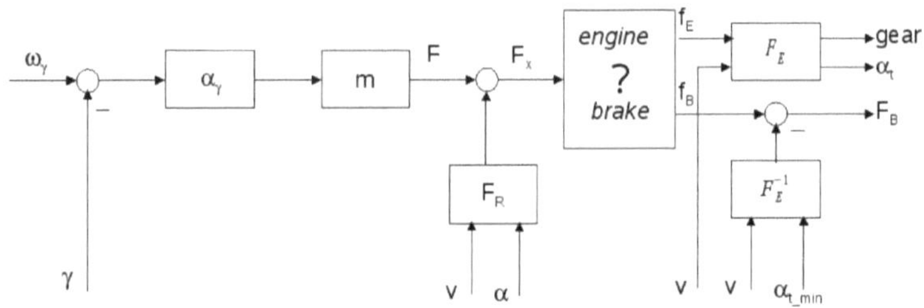

Fig. 5.3. Inverted model of the powertrain to activate the correct actuators

Figure 5.3 shows the data flow described above and is based mainly on the results of [142]. From the difference between the required acceleration ω_γ and the current acceleration γ, the force to apply on the vehicle is deduced while also taking into account the resistive force $F_R(v, \alpha)$. Then comes the verification of the inclusion of the force in the actuator range using equation 5.3.

On one hand, if the engine realizes the command, the throttle (in practice the torque request) will be determined by inverting the look-up table F_E. Moreover, if the force is higher than the engine capacity, the gear shift will be activated. On the other hand, the brakes will realize the force by looking at an internal formula to determine the corresponding torque. However, when the brakes have to realize the command, the force coming from the engine brake will be taken into account.

5.3 Possible extension to a predictive command execution by use of transfer functions

Additional information such as the road profile or the battery state (SoC) can be used to determine whether or not the gear has to be shifted. Thus a predictive adaptation of the command can be performed when additional information from the navigation are received.

On the next example in figure 5.4, the vehicle is evolving on a flat road with a supposed-constant resistive force F_R. Actually this force is not completely constant as it depends on the speed square. For the simplification of the computation, it will be set constant for the temporal extrapolation even when the acceleration is not zero. In this example, the driver requires a maximal acceleration, that is why the force coming from the combustion engine evolves with a gear shift and a change of its ratio. The gear shift will imply automatically a quasi-null acceleration while the clutch is opened, and this hypothesis is not acceptable for the driver request. Thus an acceleration boost coming from the electrical engine integrated in a hybrid vehicle can be used if the battery state is high enough.

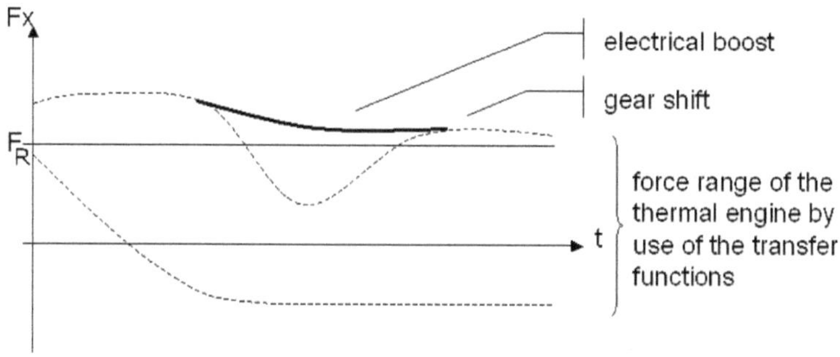

Fig. 5.4. Use of temporal models such as the electrical boost

Moreover the choice of the braking strategy can be extended as following. For low deceleration, the use of thermal engine resistance can be used but this choice is insufficient for hard braking which requires the use of the brakes. Now, with the use of the hybrid integration, the electrical engine can also be used to decelerate the vehicle as it gets a direct control to the wheels too. Thus, the following empiric law should be set to choose the braking way:

- Slow braking rate with braking demand lower than the recuperative braking capacity: use of the combustion engine resistance when the battery does not need to be loaded by use of recuperative brakes.
- Slow braking rate with higher braking command: systemic use of thermal engine.
- High braking rate: use of the brakes when the battery is full and only till the braking pads are not too warm. After that a transition rate will be used until the braking is down by the electrical engine till the brakes have an adequate temperature back.

5.4 Reactive optimization of the command

A correction of the vehicle command implies an implementation delay, when this command goes in the non-linear range of the vehicle dynamics (right side of the slip curve in figure 4.12 in Chapter II.4). This kind of correction is based on the anti-blocking skids (ABS), which can optimize the braking torque. On top of that, a traction control can be also integrated by use of that additional braking torque. Finally lateral corrections can be performed to optimize the yaw control (ESC).

5.4.1 Longitudinal correction

The braking control avoids any important slip during hard braking by lowering the braking torque adequately with an optimal wheel speed ($\dot{\theta}^{\star}$) until the slip stabilizes below 0.13 ($\lambda^\star$) (see fig. 5.5). The traction control performs the same task for the acceleration with another braking torque. Therefore both controls can be done with the same algorithm by changing only the definition of the slip.

From equation 4.7 a transfer function for the slip can be extracted by integrating the transfer functions from the wheel speed (see eq. (4.6)) and from the vehicle speed (see eq. (4.10)). The ratio of the gains describes the current state C and the final state G but it does not describe the temporal transition between the two moments:

$$\begin{cases} C_\mu = 1 - \frac{C_{\ddot{\theta}} \cdot r}{C_{\ddot{x}}} \\ C_\mu + G_\mu = 1 - \frac{(C_{\ddot{\theta}} + G_{\ddot{\theta}})^2}{C_{\ddot{\theta}} + G_{\ddot{x}}} \end{cases}$$

This slip model is then compared to the optimal slip value and the control will determine if it has to perform any corrections. If it has to modify the torque, the control will have to first define the gain G of the correction to apply to the torque:

$$\lambda^\star = G \cdot \lambda$$

The modification of the wheel speed ($\triangle\ddot{\theta}$) can be determined with a differentiation of the two equations below:

$$\begin{cases} \ddot{\theta} = \frac{\ddot{x}}{r}(1-\lambda) \\ \ddot{\theta}^\star = \frac{\ddot{x}}{r}(1-\lambda^\star) \end{cases} \longrightarrow \triangle\ddot{\theta} = \frac{\ddot{x}}{r}\lambda(1-G) \tag{5.4}$$

The correction law in terms of braking torque to be applied on the wheel can be extracted from the dynamics of equation (4.9):

$$\triangle C_{br} = \frac{r \cdot m}{k}\triangle\ddot{x} = \frac{r \cdot m}{k} \cdot \ddot{x} \cdot \lambda(1-G)$$

In terms of the transfer function, the current state C, the dynamic state D and the delay time T are defined as following:

- $C_{\triangle C_{br}} = \frac{r \cdot m}{\mu}.C_{\ddot{x}}.(1 - C_\lambda)$
- $D_{\triangle C_{br}} = \frac{r \cdot m}{\mu}.D_{\ddot{x}}.(1 - G_\lambda)$
- $T_{\triangle C_{br}} = T_{\ddot{x}}$

The transfer function of the braking correction $TF_{br^\star}$ is finally:

$$TF_{br^\star} = C_{\triangle C_{br}} + D_{\triangle C_{br}}.e^{-T_{\triangle C_{br}}} \tag{5.5}$$

Figure 5.5 presents the results of the algorithm obtained on vehicle test bench, where additional electrical motors have been connected to the wheels to emulate the road feedback. The emulated road surface has been set abruptly to icy, which has generated a high slip value. The braking control is engaged once this slip value has reached 0.2 and its control has modified the braking torque until the slip has been optimal again by 0.13 (optimum shown on the slip curve in fig. 4.12 Chapter II.4). With standard ABS systems, the braking control makes the slip converge to 0.13 anyway but with oscillations around this value as the parameters of the controller have to adapt themselves to the situation online. Here the transfer function of the braking correction is determined to be optimal prior to the intervention, because of the slip discontinuity by blocking. Hence it is possible to invert it in order to find the best braking control that avoids blocking and has a minimal converging time. But right now, the slip is only known at the current moment and when it has stabilized. The next steps will be to determine its transition between them to improve once again the action of the braking controls during the transition time.

Furthermore the traction control works only with the use of the braking units. This correction works efficiently for a short time. But for a long correction it may be better to use a modification of the engine torque also. Figure 5.6 presents these two temporal compensations. As the throttle compensation

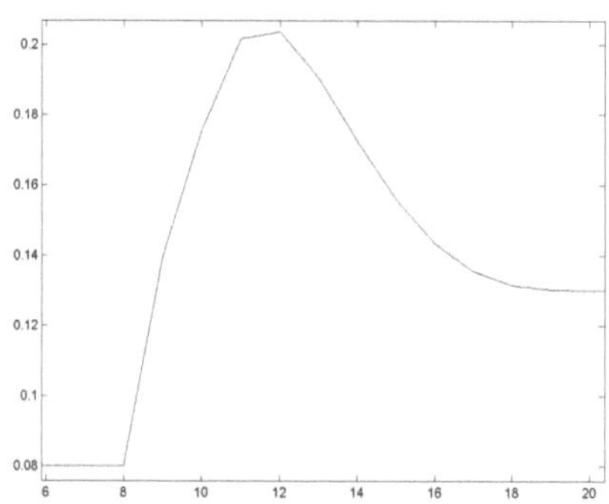

(a) Evolution of the slip due to ABS action

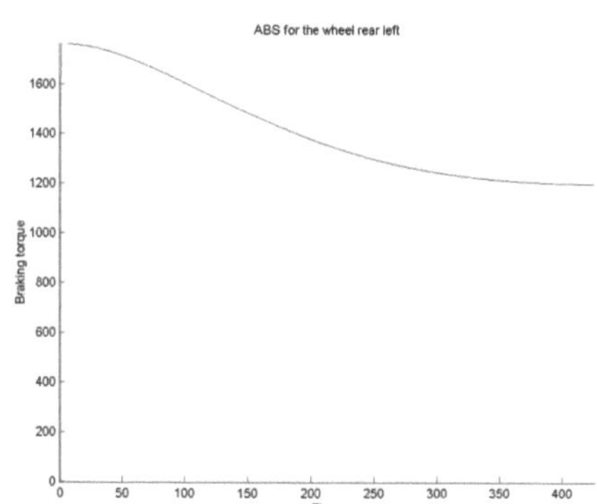

(b) Braking correction through ABS

Fig. 5.5. Correction by activating the ABS

takes more time to be released, a braking compensation is performed until the engine latency time is over. Hence the overall compensation is almost constant, but with two sources of compensation. Here the split between slow engine modification with gasoline or diesel ratio and the change of the fast modification of the combustion process will not be investigated at all but it can be seen as a third element to use in parallel.

5.4.2 Yaw rate correction with electronic stability control

The stability control is designed to detect a difference between the steering angle (δ) and the yaw rate ($\dot{\psi}$). This difference comes from an inappropriate lateral slip (β). These three elements will define the state variable (x). As β cannot be measured directly, a bicycle model as in [110] is used to estimate its dependence on the tire side slip angles α_F and α_R, which can be observed indirectly with the inertia measurement unit. The other parameters such as the distance of the center of gravity to the front axle (l_R) or the rear axle (l_F) are set as constant. The vehicle behavior as understeering or oversteering is selected in real-time depending on the lateral acceleration and the tire specifications as described in the same document:

$$\begin{cases} \alpha_F = -\beta + \delta - \frac{l_F.\dot{\psi}}{v} \xrightarrow{oversteering} \beta \approx \frac{l_R.\dot{\psi}}{v} \\ \alpha_R = -\beta + \frac{l_R.\dot{\psi}}{v} \xrightarrow{understeering} \beta \approx \delta - \frac{l_F.\dot{\psi}}{v} \end{cases}$$

Definition of the control law

The direction of the vehicle will be corrected by modifying the braking torques shortly to generate a rotation around one dedicated wheel. Let us define the

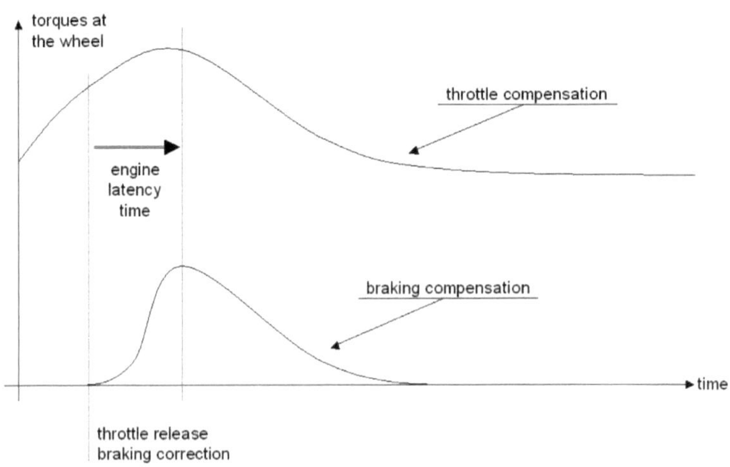

Fig. 5.6. Two-step traction correction

control variable (u) re-grouping the torques $\mathcal{T}_{i,j}$ [2] to apply on the wheels. This variable will be continuously modified by the dynamical model (f), which is based on the reduced bicycle model:

$$\dot{x} = f(x,u), x = \begin{pmatrix} \dot{\psi} \\ v \\ \beta \end{pmatrix}, u = \begin{pmatrix} \mathcal{T}_{F,R} \\ \mathcal{T}_{F,L} \\ \mathcal{T}_{R,R} \\ \mathcal{T}_{R,L} \end{pmatrix} \tag{5.6}$$

Linearized around the actual operating point:

$$\Delta \dot{x} = \frac{\partial f(x_0,u_0)}{\partial x}.\Delta x + \frac{\partial f(x_0,u_0)}{\partial u}.\Delta u$$

The control law K can be defined now to determine how to stabilize the vehicle. It is defined by $\Delta u = -K \cdot \Delta x$. The matrix ($G$) is the observer dynamic matrix in which the observer state variables are the velocity and the yaw rate:

$$\frac{\partial f(x_0,u_0)}{\partial x} - K.\frac{\partial f(x_0,u_0)}{\partial u} = G$$

Solving for K[3] gives the control law:

$$K = [\frac{\partial f(x_0,u_0)}{\partial u}]^+(\frac{\partial f(x_0,u_0)}{\partial x} - G) \tag{5.7}$$

To determine now the dynamical model (f), it is necessary to compute the Jacobian matrix [67] for the actual state. Indeed it is necessary to compute the elements (c_{ij}) which define the tire side slip constants (λ) depending on the propulsive force (F) and the road–tire friction coefficient (μ) (determined with the algorithms described in the chapter 2):

$$c_{ij} = \frac{\mu_{lat}}{\lambda_{lat,ij}} F_{ij} \approx \frac{\mu}{\lambda_{ij}} F_{ij}$$

From a reference to a braking control

Finally, the reference (x_0) for the state variable has to be defined to quantify the correction to perform. This reference is actually the current vehicle state limited depending on the speed and set after calibration. The tests have

[2] The indices (i,j) are defined by default first for the axle and then for right or left.

[3] In practice K^+ is the Moore–Penrose pseudo-inverse of the matrix K because the problem to solve is overdetermined.

been done with a passenger car (smart Roadster) by DaimlerChrysler and the results of an intervention from the stability control are shown in figure 5.7:

$$\begin{cases} \beta_{max} = 10^o - 7^o(\frac{v}{40})^2 \\ \dot{\psi}_{max} = \frac{\delta+\beta/2}{\frac{3 \cdot l_R}{2 \cdot v} + \frac{l_F}{v}} \end{cases}$$

After the acquisition of the parameters, the control law can set the braking correction to be applied.:

$$\triangle u = -K.\triangle x$$

$$\begin{pmatrix} f_{F,R} - f_{F,R_0} \\ f_{F,L} - f_{F,L_0} \\ f_{R,R} - f_{R,R_0} \\ f_{R,L} - f_{R,L_0} \end{pmatrix} = -K. \begin{pmatrix} v - v_0 \\ \beta - \beta_0 \\ \dot{\psi} - \dot{\psi}_0 \end{pmatrix}$$

The values f_{ij}, which are extracted from the dynamic model, are the current state of the vehicle. The ratio of the difference to the reference defines the gain needed for the transfer function for each braking corrections. These corrections will be transferred to the braking control described in section 5.4.1 in this chapter.

The example shown in figure 5.7 has been performed on the same vehicle test bench as for the braking controls. Once again an icy road surface has been emulated, which has made the vehicle instable in a left curve: the yaw rate has been totally different to the reference yaw rate extracted from the steering

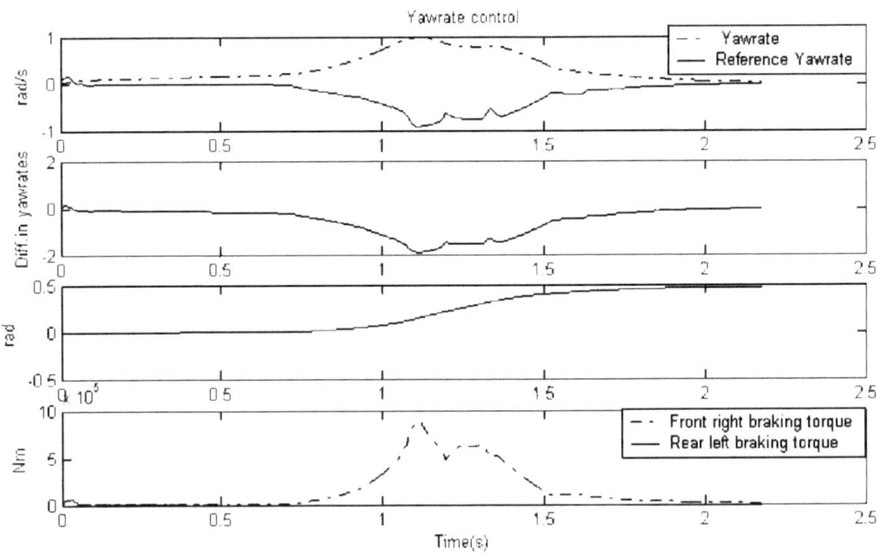

Fig. 5.7. Additional braking torque coming from the ESC to correct the incoherence in the yaw control

wheel angle. In this left curve the driver has tried to correct the vehicle sliding by oversteering by steering to the right but he could not compensate the vehicle behavior that was sliding on the opposite direction. Thus a modification of the braking force on the front right wheel has changed the position of the instantaneous center of rotation for the vehicle. After a short slope of the braking force the sliding has been compensated and finally the force has been reduced slowly to have the vehicle yaw rate correspond to the driver's request.

Part III

Virtual driver for the cooperation

1

Extended middleware for fault-tolerant architecture

Abstract. In principle, there exist two different ways to develop a virtual driver. The first historical method was based on an expert system as presented by Tascillo in [124]. However, the more complex the problem is, the more difficult it is to model and to train such a system. Thus, the upgrading of an expert system is really complex and may require a full re-design of the architecture, similar to the extension of a neural network. Therefore research has been carried on the distribution of the knowledge into several agents as in [65] to overcome the insufficiencies of the centralized approach. This methodology is based on the assumption according to which the problem to solve could be clearly split into different independent sub-problems. Here fortunately, it will be possible to use this approach, leading to having at least one agent per maneuver.

This chapter presents the middleware developed for the introduction of redundant functions like a multi-agent system. After having described the concept of software redundancy, the system management layer used for the integration of the agents will be explained. The next section will show the generic structure of the agents used as a wrapper for the safe integration of the algorithms.

1.1 Concept of software redundancy with a multi-agent system

Normally creating a multi-agent system similar to the driver cognitive model requires the use of a contract net as in [52]. Thus it will be possible to switch from normal mode (strategy and tactic) to emergency mode as in a correct driver cognition model (section 2.1 of part A). To that end, the principle has to be the same as for the human: the analysis of the time to collision as defined, for example, in [59].

However, the worst characteristic of such contract net is the singularity of the resolution process at each time. Actually, after handling with all the candidates, only the best agent will be contracted. For safety reason it is difficult to accept such simple solution, and more robust computations are required. Therefore, different versions of software will be used in parallel for algorithm redundancy as in [30].

Software failure risk can be limited by integrating redundant modules: either by replication of an agent [74] or by using parallel agents for the same task [139]. The first procedure detects problem of execution of an agent with voting algorithms [5] but it cannot detect a failure due to engineering mistake. In the second case the probability that completely different agents fail at the same time is low. That permits to make the transfer of data safe even if one agent gets into trouble. This method used in airborne systems requires two parallel teams; but here for cost reason, the automotive algorithms will only be engineered once and used in parallel to standard open agents from the robotic field.

All the data coming from the sensors are combined together into a blackboard to determine the corresponding confidence values for the control algorithms (figure 1.1). For the emergency mode, fast algorithms kept as simple as possible must be integrated. Good candidates will be the algorithms coming from the robotic as they just look how to avoid collision. The algorithms for the tactic mode will come from the automotive world, because they have the opportunity to integrate the driving rules. All the agents will either define an optimum for a maneuver or the corresponding safety envelope. If one cannot deliver all the information, it will be connected to a middle-agent, which is in charge of gathering all the dispatched data.

Two maneuvers may exist for each lane: speed and distance controls. For these tasks, at least two different agents have to deal with these maneuvers at the same time. However, parallel to the automotive agents, the robotic

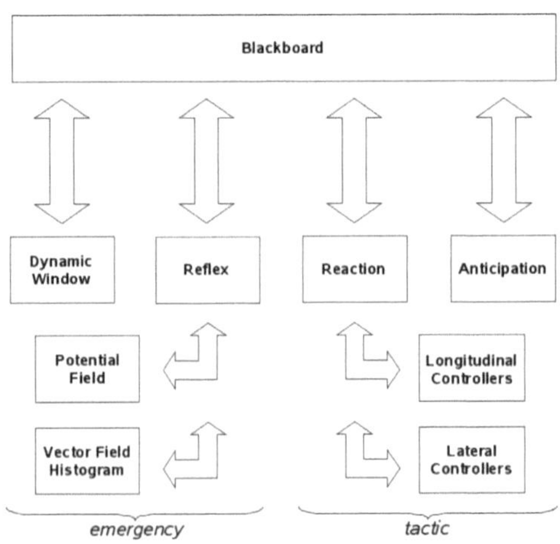

Fig. 1.1. Functional architecture of the virtual driver

agents cannot distinguish the difference between both maneuvers: these agents are only dealing with the near environment of the vehicle. Thus, the single result from the robotic agents with the different results of the automotive agents will insure the determination of the maneuver, which also lowers the dangerousness of the situation. For the two lateral lanes, only one automotive agent per maneuver is always used. Hence the robotic agents will also enlarge their monitoring scope to these lanes as soon as a danger is detected.

One automotive middle-agent will fuse longitudinal controllers (described in section III.3.2) and lateral controllers (described in section III.3.3) to define local safety envelopes and the corresponding maneuver optimums. In parallel another robotic middle-agent exists dedicated to the optima coming from the potential field agent (described in section III.2.1) and the safety envelope with the corresponding optimum from the dynamic window agent (described in section III.2.2). So, only the anticipation (described in section III.3.4) cannot be integrated twice with different algorithms. Therefore, replications are required to insure at least computation coherence for this agent or for each agent, which has no more parallel agent because of failures.

1.2 System management layer

1.2.1 Agent-based runtime environment

New generations of software architecture such as Autosar are still based on a pre-definition of the systems and their internal connections are stored in an exchanged format, used for the mappings, software configuration, etc. Several static instances are possible with these meta-modeling using UML class diagrams, for which configurations (ports, events, etc.) are performed off-line. This fundamental limitation downsizes the usability of these architectures. Indeed, no dynamic instantiation can be performed. Thus, neither software nor hardware redundancy mechanisms can be integrated correctly on the fly with parallel processors. Autosar can anyway lay the foundations of this new software architecture. Its specifications derived from OSEK-OS specifications need only to be improved on two points to solve these bottlenecks.

The control unit abstraction is also a fundamental requirement for the architecture, like the virtual functional bus by Autosar. Indeed, the communication mechanisms must not depend on the identity of the communication sources. Even if it is allowed in special cases by Autosar, it will be completely forbidden here to allow the use of substitutes. The second requirement reused here is atomicity of the data transfers. Tasks with higher priority levels (e.g., services) are able to interrupt tasks with lower priority levels. This source of data inconsistency risks needs to be taken into account during the data flow definition phase. Solutions are for example task/interrupt blocking strategy, but it skews the principle of priorities. Other solutions like sequential scheduling strategy with local copy are already used in airborne systems to insure deterministic runtime. However, such approaches deal entirely with worst case execution and they cannot optimize dynamically the runtime environment. At the level of the virtual driver, cooperative runnable placements can be defined for the agents too. Therefore, exclusive areas for inputs/outputs synchronization will be allocated especially if the costs of code overhead in the runnable prolog/epilog are insignificant with respect to the algorithms themselves like here. To sum up, the direct sender–receiver links allowed by the current architecture will be transferred to client–server links to isolate each agent and let them work in parallel with the possibility of dynamical replications.

The second important modification consists in dealing with the dynamical allocation and especially the dynamical solving process. Here the initiative of a functionality start will be shifted from the driver to the agents themselves if the agents can be activated over an electronic dongle. In practice, no function calls are required anymore, but they will be replaced by an initiative flag. All these flags together will provide information to the infrastructure administrator for the needs of replications. Because of this dynamical allocation, agents have to log-in online. To solve this second modification, a dedicated ontology is required to have the blackboard provide the correct data. Moreover, assembly connectors are also required to make the ports compatible. Otherwise, the concept of agents is still compatible with the requirements of the software components from standard architectures such as Autosar: atomic software components correspond here to agents; compositions are the same as the middle-agents. The service tasks are transferred to the infrastructure and

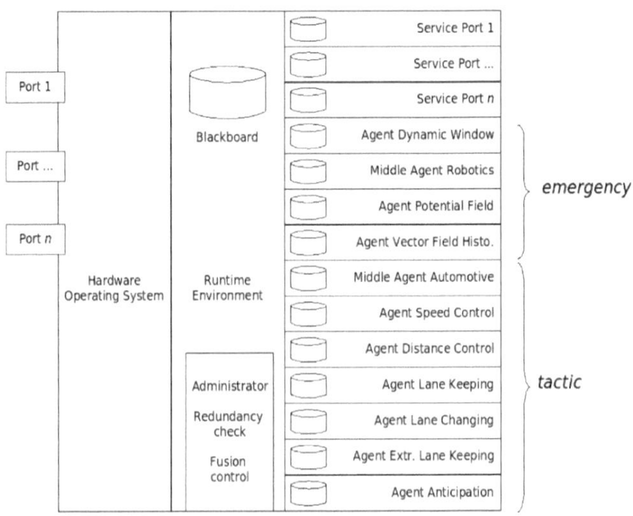

Fig. 1.2. Integration of the runtime environment with the agents

port-type converters are integrated into the mapping algorithm of the blackboard. Figure 1.2 presents the split between the different layers within the runtime environment.

1.2.2 Use of a blackboard to provide information

Blackboard management as real-time database

The main goal of a blackboard is to separate completely the knowledge from the sensors. It is possible to add functionalities between them and to avoid transfer of confidential knowledge about the sensors constellation. The standard functionalities are fusion of multiple sources of information and generation of a confidence value. Thus, a blackboard is also used on the executive level to acquire the data from the internal sensors. For example, the speed can be measured with up to four channels and their results can be combined together to avoid failures. For that either configurable Kalman filter [97] or wavelet transformation [80] can be used on demand. A log is stored for the five last cycles of each information and the corresponding measures. This history will enable the blackboard to extrapolate the missing data if any SNA [1] happens. In addition, the transfer functions obtained from the executive level are used to check the executive level state. At each time, a vehicle command is sent and some cycles later the response will come from the executive level. Thus, if the dynamics of the response is not compatible with the restriction of the maximal transfer function, failure may happen on the executive level and, thus, the inputs cannot be accepted as given.

Use of a dedicated ontology for the communication

In a static architecture, the use of a dedicated ontology is not required because the communications between modules can be linked in advance. Furthermore these connections have to be built in real time for a dynamical architecture. Thus, a common language between the agents is required to perform correctly those synchronizations. To that end, it is necessary to conceptualize the information to share [50]. Similar to human languages, standards are defined to increase the use range of the services. That is why working groups such as IEEE [94] are trying to formalize such global standards. Thus it would be possible to add extensions dedicated for new services incrementally. Most common component-based middleware use CORBA for non-critical solutions. These middleware use the task description to determine automatically the best corresponding service. For infotainment platform, some trials have already been done to get hot-plug services on central platform [104] and also on decentralized platforms [38]. Their extension for safety critical systems is difficult because the complexity of these ontologies penalizes the synchronization time with an important time overhead.

[1] Signal not available.

The simple ontology created here describes quickly the environment and it transfers the needed data to the agents. The format of the language used here derives directly from the first-order predicate calculus called knowledge interchange format [46]. KIF is intended as a language for the publication and communication of knowledge. It is intended to make the epistemological-level [84] content of a knowledge base clear to the reader, but not to support automated reasoning in that form. For each agent, the environment characterization can be realized either by the agent itself or by the blackboard. In the first case, a time intensive low-level communication is generated to bring all data from blackboard to the re-constructor by the agent. In the second case, few data need to be transferred due to high-level communication. In addition to this problem, the centralized task of the blackboard has to be kept in mind. Thus, a high-level communication will be chosen between agents and blackboard even if the wording increased rapidly. Here the blackboard will perform the environment characterization once and dispatch the relevant information to each agent. These agents will have only two possible actions terms (or verbs): they can either send a query (*get*) or reply to a query (*set*). However, the amount of nouns is much higher. Special data or full structure can be defined like: *get lane.description* or $set\ lane(\text{``}current\text{''}).(width = 3[m]\ and\ mark(\text{``}left\text{''}) = \text{``}dotted\ line\text{''})$. If any information is not available, a not-defined word will be used. Otherwise a couple value–confidence will be provided. For objects, reliable properties can be received with different indexing. For the example in fig. 1.3, the next sentences are equivalent:

$$get\ lane(\text{``}left\text{''}).(object(position = \text{``}ahead\text{''}))$$

$$get\ lane(\text{``}left\text{''}).(object(distance > 0))$$

$$get\ object(id = 3)$$

$$get\ object(closest)$$

Some information has low-level meaning such as the *id* identification. However, some other requires a prior analysis from the blackboard such as *closest*. The second case is especially used for the road analysis. Indeed, the analysis for crossing section is much more complex than vehicle characterization and the intersection topology needs to be integrated here as: $lane(intersection) \longrightarrow false$ or $lane(intersection(type = \text{``}K1\text{''}))$.

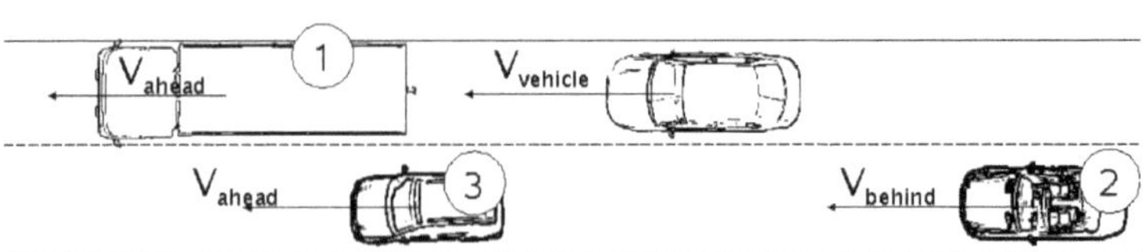

Fig. 1.3. Example of equivalent indexes with ontology

1.2.3 Redundant management of the agents

An administrator is integrated in the runtime environment with the aim to look which agents have taken initiative to control a maneuver. This agent monitoring consists first in the error handling by isolating the agents. Secondly, as the different robotic agents cannot describe the maneuvers by themselves, the clustering of the results has to determine which agents are used redundantly. For the other single agents, replications will be asked directly. Figure 1.4(b) shows the data flow of this runtime environment.

In the same figure, the safety envelope, which will be defined by the agents, is described with the problem of coherences between the outputs of the different agents. For example, different agents will provide different optima for a same maneuver, and different agents will provide different limits of the steering range and the acceleration/braking range. All these agents provide anyway compatible results that have to be combined together.

The problem is to know how to fuse all the data given by the agents of the virtual driver. More precisely, it can be divided into two sub-problems: a

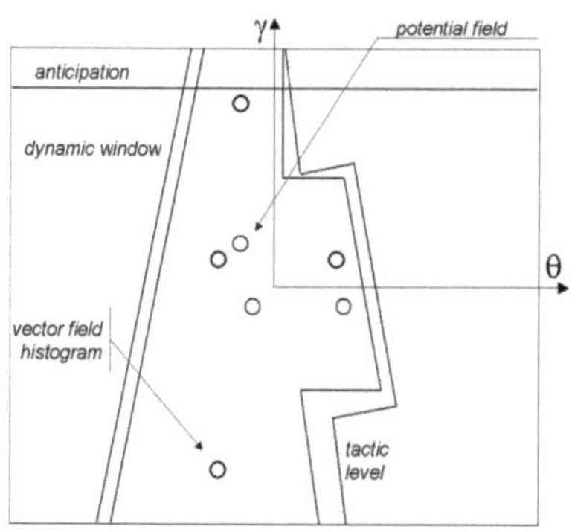

(a) Fusion of different solutions

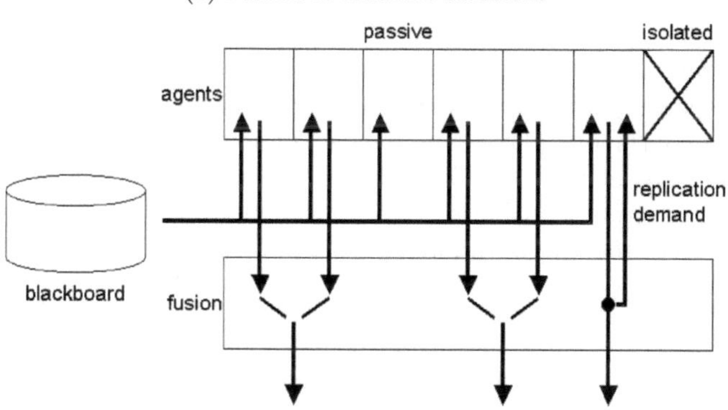

(b) Administration of software redundancy

Fig. 1.4. Principle of software redundancy and alternative with replication problem about the safety envelope of the maneuvers and a problem about the optimal commands. So, the analysis to fuse the different information of the agents will be different for these two kinds of data.

Clustering methodologies

Clustering [97] is used to group agents with similar objectives or data. In a multi-agent system, clustering is viewed as the search of individual agents having the goal of finding similar information. Agents within a cluster coordinate by combining their local views to allow each member to search a broader range of neighbors for better matches.

K-means clustering

The technique [63] considers initially each data as a cluster. It chooses the K cluster centers, called centroids, and compares each of the K elements in the data set to each centroid, placing each element in the cluster to which it is the closest. Then it re-calculates centroids based on the resulting cluster memberships and repeats this process.

Hierarchical clustering algorithms

This method [107] consists in considering initially each data as a cluster, building a table of distance or similarity between every pair of clusters in the data set. With the use of this table, a series of partitions is created by splitting or combining clusters based on the closest (most similar) pair of clusters. Then, the new cluster and each of the old clusters are considered with the table of similarity and then the combination is made. These two steps are applied until all elements are in adapted clusters.

Clustering as a combination of Gaussians

This technique of clustering [42] considers that each cluster can be mathematically represented by a parametric distribution such as a Gaussian distribution for a continuous phenomenon or a Poisson distribution for a discrete phenomenon. So, in this method, the entire data set can be modeled by a combination of these distributions. And by comparing the coefficient of the distributions, clusters can be found with computation of the new distributions.

Fuzzy C-means clustering

The fuzzy C-means clustering [43] is a method of clustering which allows data belonging to several clusters. It permits to recognize pattern with a minimization of an objective function. More precisely, this function is based on the definition of the degree of membership of an element in a defined cluster and on the similarity between the element and the center of the defined cluster.

Then, an iterative optimization of the considered function is computed until the termination criterion based on the absolute difference between the degrees of membership of each element at the step k of the iteration and the degrees of membership of each element at the step $k+1$ of the iteration are respected. This criterion is defined arbitrarily in advance with the wished precision of the clustering.

Determination of the clusters

Let us recall the main difference between robotic agents and automotive agents: robotic agents are not maneuver oriented but safety oriented. Thus, the clustering is a fusion of two different behaviors. In such cases, optima from robotic agents can correspond to different automotive maneuvers at the same time. So, they can correspond to more than one cluster. Indeed, a fuzzy

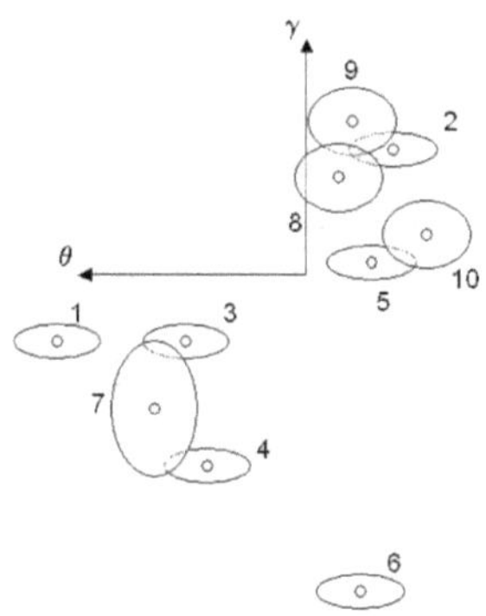

(a) Initial position of the optimum

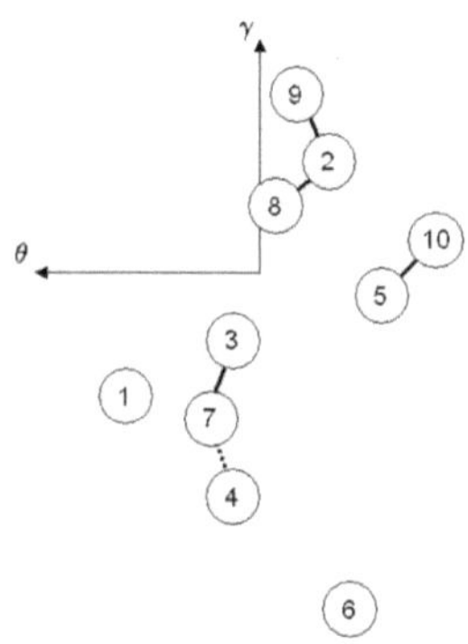

(b) Primary clusters from the graph of connections

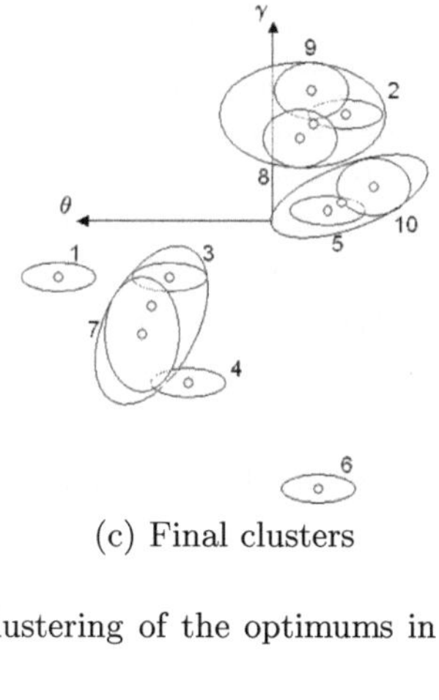

(c) Final clusters

Fig. 1.5. Three step of clustering of the optimums in the (acceleration–steering) space

C-means clustering seems to be the most adapted methodology here. The decomposition into clusters can be accelerated by using a modified hierarchical clustering algorithm, which will have at the end the same rough behavior.

If the results of an agent have not been updated the last data will be reused and extrapolated linearly. In this case the confidence value will be degraded and the size of the uncertainty range increased. After four next missing samplings, the output of the agent will be automatically disabled because a further extrapolation would not have any meaning anymore and the agents will be rebooted. Moreover if an agent has started its computation before the coming of a potential danger, it will be restarted automatically. Otherwise its output, which cannot take into account the new requirements, will have a negative effect on the consolidation.

Initially each optimum represents a single cluster. Then, the table of distances is used conjointly with the table of confidence values to determine the acceptance of fusion of prior clusters. Actually, the confidence value is used to define the surrounding envelope of uncertainty of the optimum. This envelope will be looking like an ellipse as the uncertainty of the computation is higher for longitudinal controllers than for lateral controllers. The acceptance of fusion will be positive as soon as the envelopes intersect. With this first step, a graph of connection can be fully built and the primary clusters extracted. Figure 1.5 shows the three steps of the clustering algorithms for the optimums of the maneuvers. The automotive optimums and the robotic optimums are represented with dots and the center of the resulting clusters are circled with the confidence area.

Furthermore the full clustering has to be based on the automotive optimums, as they are maneuver oriented. The modification of the standard hierarchical clustering approach is to give here the possibility for the robotic agents to belong to several clusters. Hence, degrees of membership can be defined for clusters depending on the similarities, which here corresponds the distance to their new center. Thus, even if one robotic optimum intersects different automotive optima, they will not be fused all together. On the former example, the connection (4,7) will be broken to avoid this amalgam.

Iterations will be performed until the degrees of membership of each optimum are stabilized. At each cycle time, the optimums and the center of the partial clusters that have the best connections will be clustered together. From this new cluster, a center will be defined by weighting the positions of the different optimums with their corresponding confidence value.

Determination of the local safety envelopes

Once the clusters of optima are defined, their safety envelopes can be fused together. Agents with higher confidence rate will consume more area of their safety envelopes. This reasoning has been chosen instead of a more conventional probabilistic approach, which will define a framed range around the boundaries of the safety envelope. Indeed, the algorithm has to stay conservative: it is better in most cases to decrease the driving possibilities as to leave too much freedom.

After forming the common range, which can be seen as the core of the safety envelope fusion, this area will be extended on the basis of the confidence levels of the agents. At first, the intersecting points (in the figure 1.6) and the common points (in the same figure) of the two decreased envelopes are detected to generate the common range. Then, this range will be extended into each direction. First the extension from the intersection point on the boundaries can be performed relative to the confidence level as direct ratio of the extension. Then, the diffusion of the common points can be performed. However, the direction of the diffusion is not so easily defined here. Actually all possibilities can be similar to the two cases below:

- The amount of apexes, which are not already attracting an intersection point, is the same as the amount of common points: each point is attracted by its closest corresponding point as in (figure 1.7).

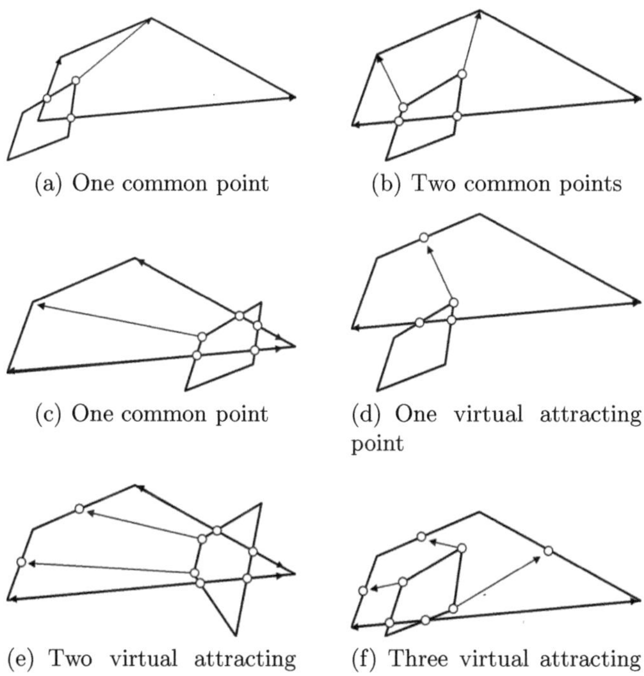

(a) One common point
(b) Two common points
(c) One common point
(d) One virtual attracting point
(e) Two virtual attracting points
(f) Three virtual attracting points

Fig. 1.6. Six standard examples of attraction

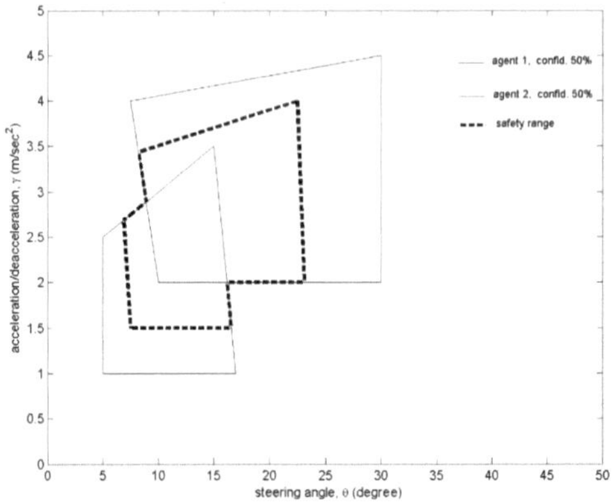

(a) One common point

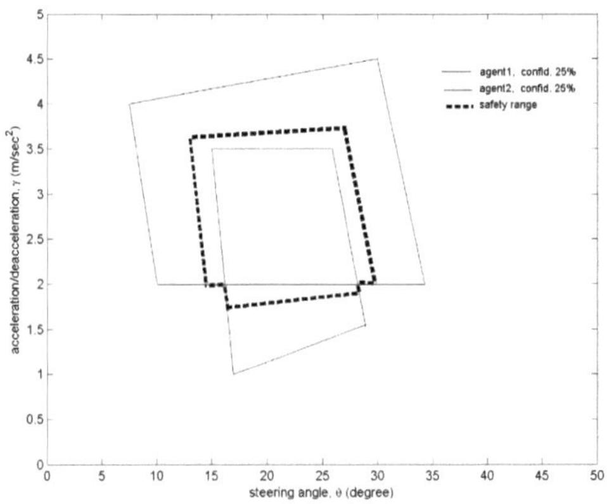

(b) Two common points

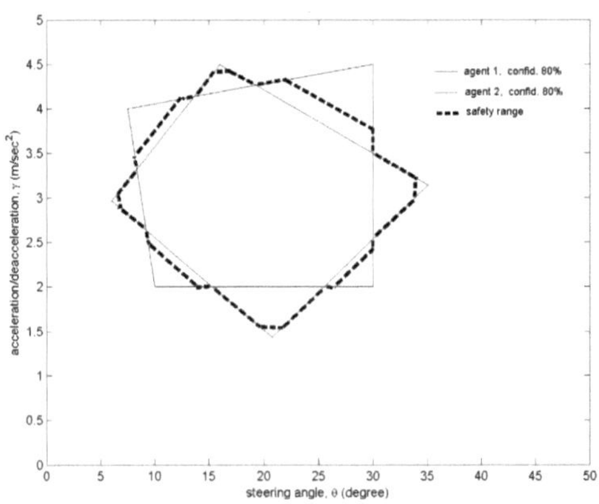

(c) Without common point

Fig. 1.7. Three final examples of safety envelope fusion

- The amount of these apexes is the amount of common points plus one: the middle of the free edges will be used as virtual attracting points (in the figure 1.6).

1.3 Integration of fail-tolerant agents

A generic class is defined to create agents. Each agent can provide information relative to a maneuver: optimum and/or safety envelope. Agents, which cannot deliver full information, will be linked to a middle-agent. A middle-agent is a sub-class of standard agent, because it has also the possibility to deal with partial agents to fuse their data.

1.3.1 Structure of an agent

An agent class is actually just a wrapper written in C++ to encapsulate algorithms written in C within a dedicated workspace. Every agent has the same BDI [137] architecture. The Beliefs are a local copy of the relevant information performed to avoid data corruption. The Desires which are defined in advance are the data to compute. And the Intentions corresponding to the way to achieve the desires, correspond to the algorithm itself.

Functional architecture

Besides this logical architecture, another functional architecture is defined. Three layers are used to define the agent. The lowest level (ISO/OSI 2) integrates the driver for the transfer of a data with the blackboard by use of the KIF/KQML language. The interaction layer (ISO/OSI 4) uses the lower layer for the full synchronization of the query/reply from the last level (ISO/OSI 7). Thus level integrates also the ontology. The last level integrates the pattern to take the initiative and the algorithm of resolution. On top of that, a proxy can change from passive mode to active mode, where multiplication of the agent is required. As the pattern to take the initiative has to be clearly respected, it will automatically be prelisted through the proxy to ensure the correctness of its computation (figure 1.8).

The N-voting algorithm will be integrated here and not directly into the administrator. With this approach, multiple instance of a middle-agent will have repercussions automatically on multiple instantiations of the connected agents. Thus, a real transparency of the computation chain tool is possible during instantiations.

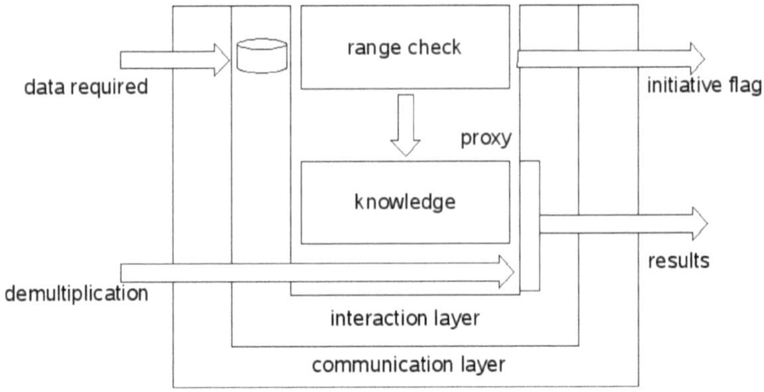

Fig. 1.8. Functional architecture of an agent

Practical integration

In practice, the agent has standard functions which are generically defined and they will only be inherited and modified over polymorphism:

1. *Connection*: Definition of the local database and the desires of the agent. The administrator or the middle-agents will store which inputs are connected to which data
2. *Initialization*: Initialization of the memory space dedicated to the agent. For example, look-up tables will be stored on static memory pages for later use. The initialization of a middle-agent is just transferring the signal to the connected agents and the required data.
3. *Getting*: Grabbing all the data. This impulse will provoke the synchronization of the data.
4. *Initiative*: The agent checks if it can take the initiative. The result may be Boolean or even an extension with a level of matching of the pattern for the probabilistic extension.
5. *Step*: The agent takes the initiative and it will determine the desires, which it has had previously defined. Prolog and epilog will be possible if the agent requires it.
6. *Termination*: De-allocation of the memory used by the agent. It looks almost like a destructor. For the integrated C code as algorithm, this termination is almost void because no dynamical memory allocation is allowed [87].

1.3.2 Redundant computation

The internal redundancy is dedicated to agents (e.g., anticipation task), without parallel agents. During this computation time, the bottleneck will be the proxy in charge of the internal synchronization of the replicates. If any failure comes from its part, the full agent will fail.

Active or passive replication

The approach developed the first time by Avizienis in [5] and improved for example by Ralph Deters in [37] for fault tolerance is based on the notion of replication. Indeed, all members of the replication group receive the same message. The group of replications also permits to detect the case in which an agent of the replication group fails. Then, the method is to treat the group as a single object behind one proxy.

In an active replication, the replicants compute their own results following a concurrent method. Thus, all the replicants of a group are active at every instant. Furthermore, the coherence of the results is checked by comparing at each moment the different computed results given by the replicants of the considered group. So, as a failure or a fault can be easily located, the replicant that is the cause of this failure is not taken into account for the computation of a global result and its process can be finally killed. Then, a check with all the different results given by the replicants is performed to obtain a result for the replicant group.

In a passive replication, only one replicant of the replicant group is active and the others are passive. The active replicant is changed every sample time. Then, the active replicant checks its result with the history to ensure the coherence of its result and to maintain the coherence of the group following this method: the results given by the active replicant are checked with the evolution of the previous temporal results computed by the other replicants that are at that moment sleeping. If the active replicant results are incoherent, it will be replaced by a sleeping replicant. Then, the correct results of the active replicant are returned. This method implicates less memory and a low time overhead; however, it also inserts this following problem. An agent can fail without being detected and then, the following agent will be detected as failed agent. So, a controller has to be implemented to avoid this situation.

The determination of the number of replicants

First of all, it is interesting to define the criticality of an agent, that means its rate of activity and its domain of application. Then, the criticality of an agent allows to define its importance and influence into the multi-agent system and to choose the repartition of the resources between the different agents by creating replicants. And so, when an agent has a high criticality, it will have at disposal a relative high domain of resources and a relative high number of replicants. More precisely, for the safety range, only the a priori knowledge can be used to determine the criticality of the agents. But, for the comfort range, the a priori knowledge and also the evolving knowledge are used to determine the criticality of the agents.

The a priori knowledge

The first step to evaluate the number of needed replicants for each agent is to develop a first appreciation by analyzing the different criterions which characterize the considered multi-agent system. Although each agent is autonomous in a multi-agent system, an agent has not always all the required competences and resources. So it needs the help of middle-agents and these links have to be taken into account to evaluate the number of needed replicants for each agent in a considered multi-agent system. The use of the interdependence net permits to have a representation of the multi-agent system and to have the different weights in step of these principal criteria:

- The size of messages between the different groups
- The number of messages between the different groups
- The kind and the sequence of the messages

The method of mathematical evaluation

As mentioned previously, it is possible to evaluate the number (N) of replicants for an agent with the resource at disposal (RD), the minimal number of replicants (MN), the criticality of this agent (CA) and the sum of the criticalities of the other agents (SCA) by following this mathematical formula:

$$N = |MN + \frac{CA \cdot RD}{SCA}|$$

The synthesis of the results between the replicants

In this case, the different replicants have to give the same results and so, a simple check to find the errors is only needed. For this approach, if a group of replicants as one object or an agent that is not a replicant needs to communicate with another replicants group, it has to know the number of entities in the replicants groups and so the number of messages to send, and the entities that have to receive the message.

Moreover, there is a problem with the proxy: there is an introduction of a single point of failure problem. To decrease the influence of this single point, the agents have to have adjustable roles: replicant as a single agent, as a member of a replicants group and as a communications proxy for a replicants group. If an agent as a proxy fails, the replicants detect this and select one of them to adjust its role and take over as a proxy. The data proxy is used to address the communications between the replicants and the environment. It ensures that all replicants receive the same percept of the environment and handle results synthesis. It introduces a single point whose influence can be deleted with the introduction of agents that have four adjustable roles (a single agent, a member of a replicants group and a communications proxy and a data proxy for a replicants group).

2

Agents derived from the robotic field

Abstract. Different well-proven algorithms coming from the robotic field can be integrated in this place. On one hand a function based on a potential field method can define an optimum of command in order to avoid the sources of danger. On the other hand an extended dynamic window can be used to set roughly the limits of the safety envelope. However, these algorithms, which are geometrics oriented, do not verify whether the path planned is physically feasible or not. Hence such algorithms are normally supposed to be used to determine a possible first path planning, which in turn has to be verified.

2.1 Potential field approach

The agent based on a potential field method such as in [72] computes firstly a rejection force for each object and for the street and secondly an attraction force for the destination to reach. The sum of these forces defines the acceleration and the direction of the vehicle motion. Therefore this method can only deliver an optimum for the maneuver to stay on a lane (figure 2.1).

2.1.1 Rejection forces

The rejection forces are directed from the objects to the vehicle. The norm is dependent on the type of object (A), dependent on its angle δ and the inverse of the distance d in a power of n. This value n is an integer between 2 and 5 depending on the type of object. To the end, the force gets a limit in order to avoid any divergence to the infinity:

$$||\mathbf{F}||_2 = min(\frac{A}{d^n}\delta_\theta, F_{max}) \tag{2.1}$$

The orientation (δ_θ) of the vehicles[1] relative to the ego vehicle is also taken into account. This permits to match a heterogeneous 2D pattern of the

[1] The orientation is not taken into account for pedestrians.

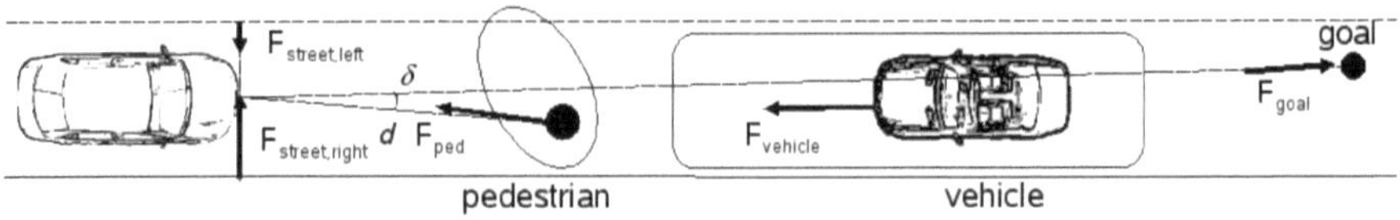

Fig. 2.1. Forces on the vehicle

field around the object. Otherwise a vehicle parallel to the ego vehicle will generate the same rejection force than a vehicle ahead. The pattern is defined for a vehicle like a drag depending on the ego speed to have the distance of the ego vehicle respecting roughly the safety distances (figure 2.2).

As long as the objects behind the ego vehicle are not dangerous, no action is required from the vehicle in term of acceleration. A dangerous situation will come only if the relative speed of the object is too important for the current distance. In this case a rejection force will be also computed for these objects in order to avoid risks of collision, which will generate an acceleration of the ego vehicle.

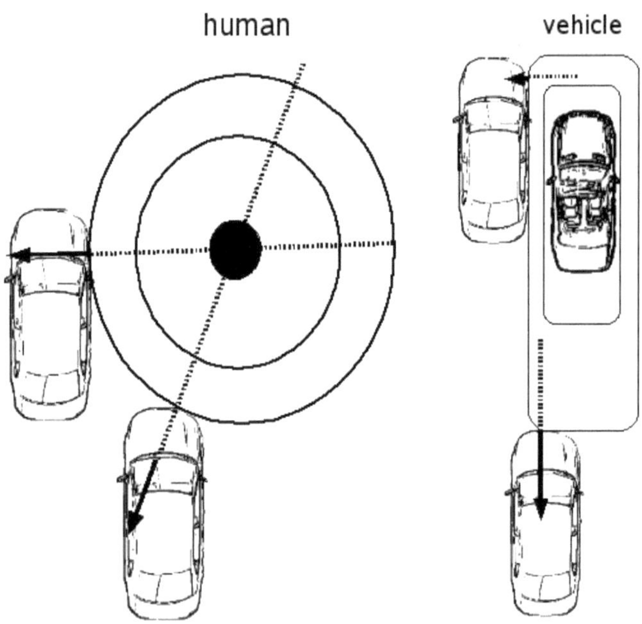

(a) Oval pattern for the potential field of a pedestrian

(b) Rectangular pattern for the potential field of a vehicle

Fig. 2.2. Pattern due to the orientation of the objects and variations of lateral displacement of the ego vehicle

2.1.2 Lane keeping

Moreover the street is defined dynamically as two objects parallel to the vehicle on the lane. The summation of their rejection forces creates a tunnel of force whose minimum corresponds to the middle of the lane (fig. 2.3). Finally these rejection forces are limited close to road borders to avoid having a lateral force tending to the infinity, which means an abrupt steering wheel. Indeed such steering modification at high speed can be dangerous for oversteering vehicles when they reach the critical speed limit for the steering sensitivity.

2.1.3 Temporary destination setting

The temporary destination is set dynamically after the last object ahead the ego vehicle on the same lane in order to have the vehicle with the demanded speed or the maximal speed allowed. To do this, a controller is integrated to modify the distance of the goal in a limited range to have the vehicle speed reach this demand.

2.1.4 Resulting acceleration

The summation of the forces will provide the acceleration on the plane:

$$m \cdot \gamma = \overrightarrow{F_{goal}} - \sum_i \overrightarrow{F_{object,i}} \tag{2.2}$$

The acceleration will be split into longitudinal acceleration γ_L and lateral acceleration γ_l. The longitudinal acceleration will be set as the vehicle acceleration and the lateral acceleration will set directly as the steering angle θ if it is physically possible, by using the Ackerman model presented before in

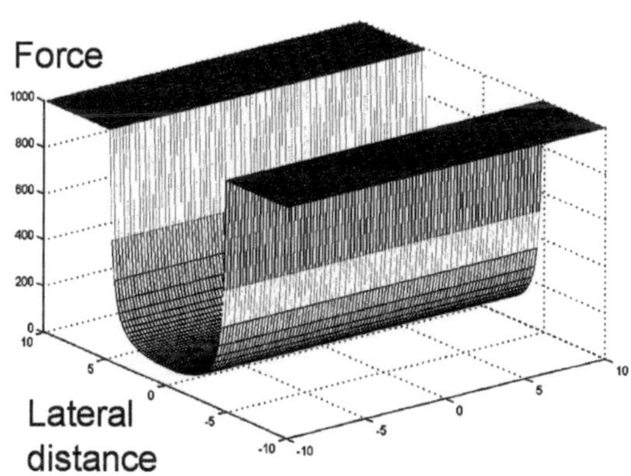

Fig. 2.3. Generation of a potential tunnel for the lane keeping

figure 4.7 (Chapter II.4) and defined in the equation 2.3 again using the standard vehicle parameters:

$$\theta = \frac{l}{R} + \frac{m}{l}(\frac{l_r}{c_{sf}} - \frac{l_f}{c_{sr}})\gamma_l \tag{2.3}$$

In the example in figure 2.4 the ego vehicle stays at a distance of 2 m behind another vehicle, which has a constant acceleration of 3.5 m/s^2. The ego vehicle gets a lower acceleration until it reaches the safety distance of 8 m. At this moment the ego vehicle will also have the same acceleration as the vehicle in front.

2.1.5 Resulting problem

However, if the resulting force is too important and corresponds to a hard braking ($\gamma < -4$ m/s^2), the lanes delimitation will be modified automatically in order to give the possibility to change the lane to the vehicle as in figure 2.5. First the left lane will be disabled if the marking is adapted and after that the right lane will be disabled if it is possible. The choice of the new lane will depend directly on the lateral position of both vehicles, which influences the equation 2.2 presented before. Indeed the lateral forces because of the road will be lower on the modified border(s), which will produce a variation of the lateral acceleration of the vehicle.

For the same reason if the vehicle is aligned with an object and the destination, the resulting force will be only longitudinal. This local minimum has to be avoided by adding in this case a lateral force to the left to always get the acceleration stable. Otherwise it should be possible to have the ego vehicle staying between two objects and colliding with them because the situation would not have generated lateral forces.

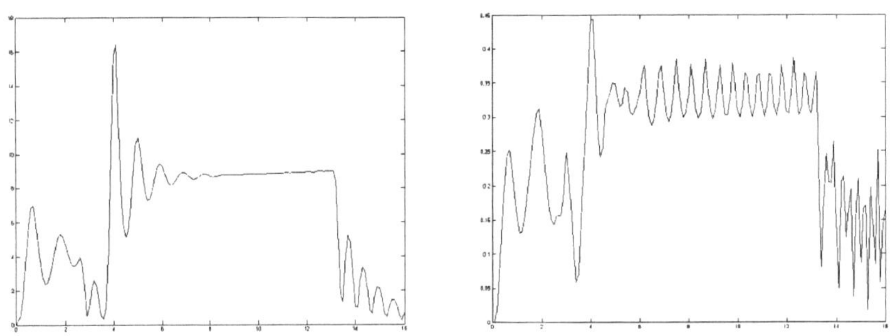

(a) Distance to the vehicle in front

(b) Acceleration of the ego vehicle

Fig. 2.4. Example of distance control by use of potential field method

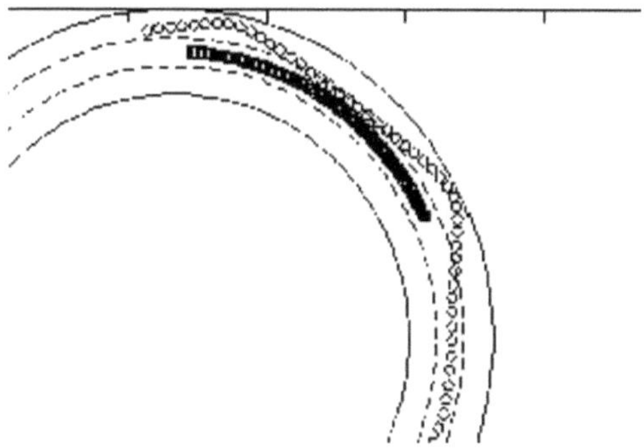

(a) Lane changing generated by an important braking rate

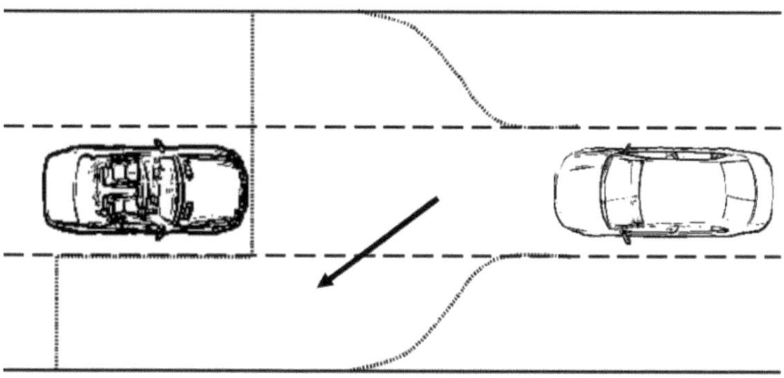

(b) Corresponding lane changing process

Fig. 2.5. Integration of the lane changing possibility

2.2 Modified dynamic window

The second agent is optimized for the pedestrian avoidance, where the computation time allowed is very short. It will focus on the objects/pedestrian ahead as 70% of the accidents happen on the front. Moreover 94% are moving pedestrians; this situation requires the integration of a dynamic extrapolation and a fusion of those extrapolations.

This function is based on a modified dynamic window. Normally a dynamic window downsizes the computation of the safe plane (speed, rotation) to the safe reachable range within the next time step. Contrary to standard method such as in [4], the computation of the safe command is not defined with a constant curve radius but with a constant steering angle. In this case the vehicle motion depends on the road–tire friction coefficient and on the longitudinal and lateral slips. Therefore a single track vehicle model is integrated to extrapolate the paths instead of defining a path as a constant curve.

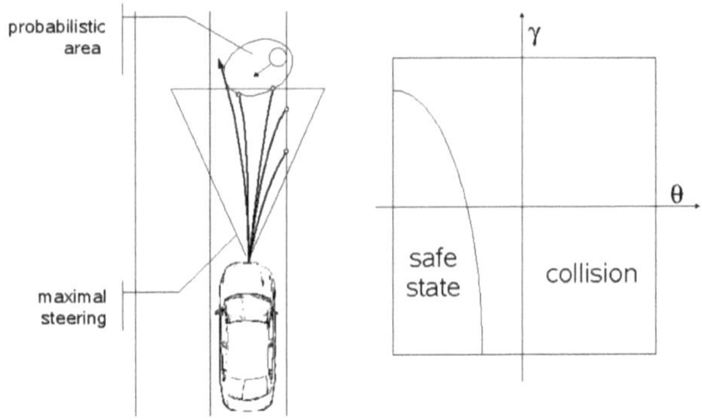

Fig. 2.6. Generation of a safety envelope

The computation of the safety envelope is based on the fusion of two needs (figure 2.6). On one hand the vehicle has to stay on the road by getting the safest steering angle. On the other hand the commands leading to a contact with the object have to be rejected. The safe commands will be acceptable only if they pass both requirements.

2.2.1 Road monitoring

The maximal steering angle will be defined when the vehicle provokes only one contact with the border of the road and its steering angle is modified with the maximal steering rate as shown in figure 2.7. That is why it is necessary to find the steering angle for which the distance between vehicle and road border is null on the global minima. If the acceleration is too strong for the driver, the road monitoring will be extended to the other lanes.

The distance (d) to the lane is extracted from the parametrical equations of the road and vehicle trajectories ($\mathcal{T}$); unfortunately θ_{max} can only be solved numerically by using Taylor approximations. This method approximates the solution of d to a polynomial to the degree 3 with a precision of 15%, which is enough for angles up to 50°:

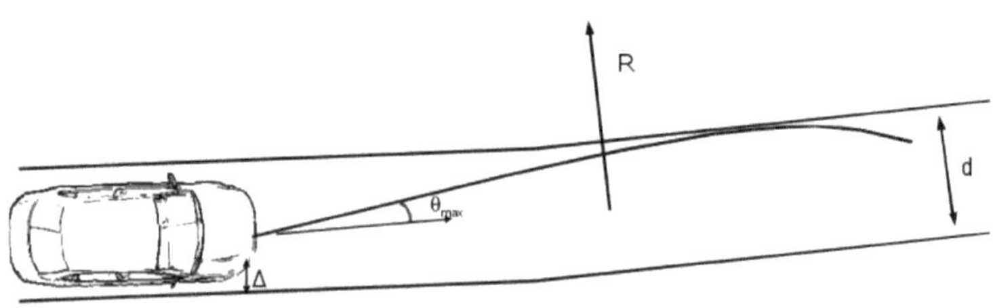

Fig. 2.7. Definition of the last safe path

$$\begin{cases} \mathcal{T}_{r,x}(\tau_r) = (R - \frac{d}{2})sin(\tau_R) \\ \mathcal{T}_{r,y}(\tau_r) = (R - \frac{d}{2})cos(\tau_R) - \Delta + R + \frac{d}{2} \end{cases}$$

$$\begin{cases} \mathcal{T}_{v,x}(\tau_r) = -\dot{\theta}_{max} \cdot V(sin(\theta_0 - \dot{\theta}_{max} \cdot \tau_v) - sin(\theta_0)) \\ \mathcal{T}_{v,y}(\tau_r) = \dot{\theta}_{max} \cdot V(cos(\theta_0 - \dot{\theta}_{max} \cdot \tau_v) - cos(\theta_0)) \end{cases}$$

$$d = \sqrt{(\mathcal{T}_{v,x} - \mathcal{T}_{r,x})^2 + (\mathcal{T}_{v,y} - \mathcal{T}_{r,y})^2} \tag{2.4}$$

2.2.2 Object monitoring

Two maneuvers are possible to avoid an obstacle. If the braking capacity is enough to stop the vehicle before the collision, the vehicle will be able to stay on its trajectory. Otherwise it will have to change its trajectory either to the right or to the left. This choice can be done by taking into account the position of the other obstacles, or the direction of the most important obstacle.

Therefore it is necessary to define the maximal steering angles as shown in figure 2.8. Between these two angles the acceleration has to be limited in order to get the vehicle stopped. In this case the road–tire friction coefficient μ is taken into account to determine the maximal longitudinal braking capacity and the capacity of lateral translation.

Unfortunately as this algorithm is used for robots, the sliding due to the tires is not taken into account. Therefore the theoretical circle path with the Ackerman radius cannot be used anymore. Here a parameter g will be introduced to define the difference between the theoretical model and the real measures:

$$g = \frac{\theta_{measure}}{\theta_{Ackerman}} \tag{2.5}$$

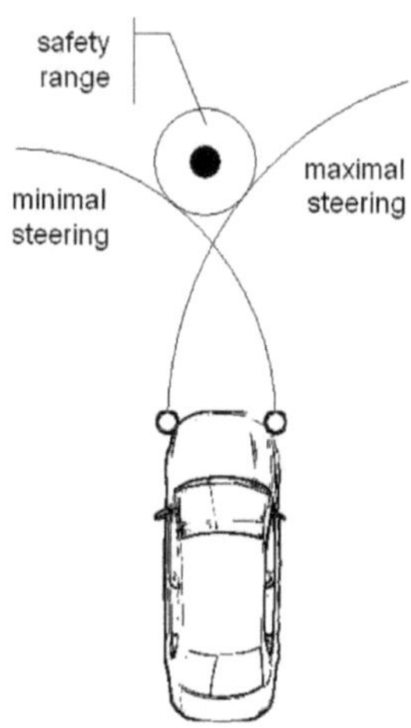

Fig. 2.8. Steering range for an avoidance maneuver

The problem is that this factor depends on three different parameters: the initial speed, the deceleration and the position of the object. It depends also on the friction parameter (μ), but this parameter is seen here as a constant. The idea now is to analyze the development of g when two parameters remain constant and the last one varies. The most impressive results are obtained when the development of g is presented for object position and deceleration in dependence on the initial speed v_{init} as in figure 2.9. These data have been simulated by using a vehicle model, which has been calibrated for a passenger car. Next some non-critical results have been validated by using that passenger car.

In figure 2.9, the variable g is described for each speed. Three ranges can be extracted: the vehicle can stop before colliding with the obstacle (the lowest range in black), the vehicle can steer enough to overtake but it cannot brake enough anymore to stop the vehicle (in white) and in the last range the vehicle can neither brake nor steer enough to avoid the collision.

However, the cases representing g with positive collision avoidance now give the possibility to develop a function $h_\gamma(v_{init})$ that is able to calculate roughly g with respect to v_{init} and therefore the correction coefficient to

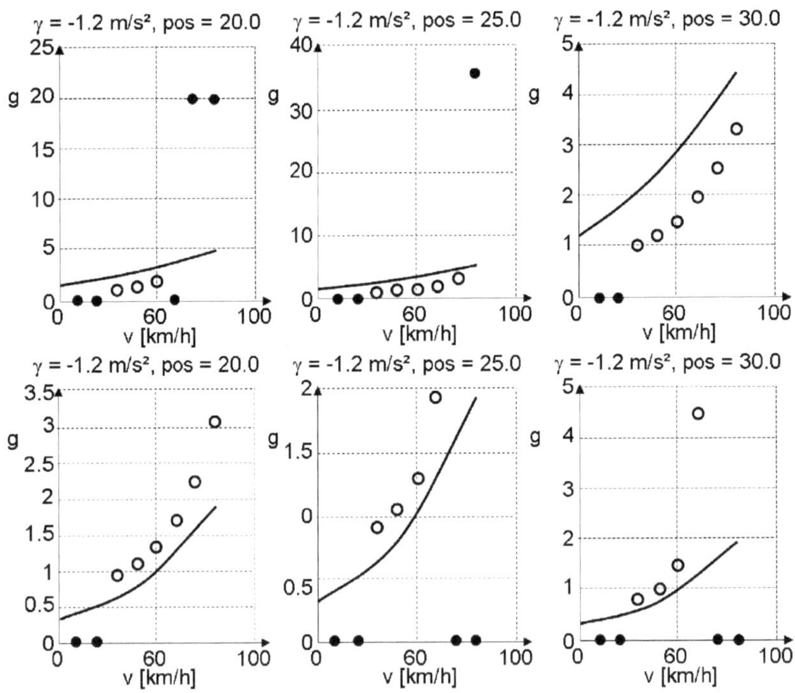

Fig. 2.9. Development of g for object position and deceleration in dependence of v_{init}

apply to the theoretical computation. This interpolation can be realized, for example, with the help of the Lagrange polynomials. This function is determined once for each deceleration and for all object positions whose resulting radius r_t has the same sign. A further condition is to add to the computations a security factor seen as translated curve in figure 2.9. This means for the function computing g for all positions the final calculated factor has to be always smaller than the factor determined with the simulations.

2.2.3 Fusion of the two sub-modules

Once the limits of the steering range (figure 2.10(a)) and the object avoidance (figure 2.10(b)) are defined, a fusion of both results will provide the definitive

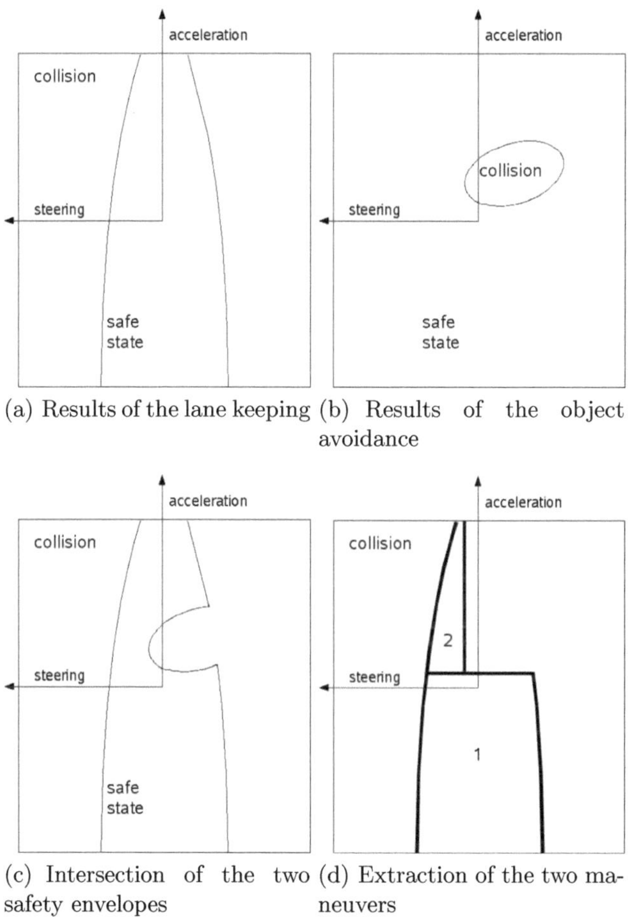

(a) Results of the lane keeping (b) Results of the object avoidance

(c) Intersection of the two safety envelopes (d) Extraction of the two maneuvers

Fig. 2.10. Different steps to generate the safety envelope from the dynamic window safety envelope. The so-obtained safety envelope comports two ranges in which objects and road boarders cannot be separated easily (figure 2.10(c)). The second step consists in separating the safe range into two zones: a range beneath the maximal acceleration to get the vehicle stopped ahead the object, and an upper range where a lane changing is required. As it can be seen in the last figure, the range above the object avoidance is also disabled for safety reason even if it is theoretically possible at this computation time to avoid the object by accelerating. The lower optimal point in figure 2.10(d) corresponds to stay behind the object, to brake and to hold the vehicle on its current lane, whereas the second maneuver will pass the object with a lane changing.

3

Tactic agent for speedway/highway

Abstract. Each time, the safety envelope is defined by the vehicle capacities. This information, which is continuously coming from the executive level, sets these limits directly. Furthermore the safety envelope can also be limited by actions dealing with the environment. Actually the vehicle capacities limit the safety envelope, but in dangerous cases, the safety envelope will be further limited by the limits of the safe maneuvers.

In this chapter, the agents native from the automotive world are presented. First the reactive agents dealing with longitudinal control such as speed and distance controls will be explained. After that the description will continue with the lateral controls such as lane keeping or lane changing. Toward the end the anticipatory action to prevent over-speed will be presented.

3.1 Fusion of reactive and anticipatory action

Like in the human cognitive model two elements continuously conflict together. On one hand an agent has to react to the local environment of the vehicle in order to define the instantaneous safety envelope. This bottom-up approach (corresponding to triggered event) determines the feasible maneuvers depending on their results, which should not be hazardous. And on the other hand an agent has to anticipate problems coming ahead. The pre-compensation by this agent has to modify the safety envelope (as shown in the figure 3.1) by downsizing the maximal acceleration needed in order to always have an adequate speed for the next moment. This top-down approach (corresponding to standard actions) looks at the safety goal to extract which maneuvers can be performed.

Each maneuver has to be limited by the maximal preventive acceleration as depicted in figure 3.1. If any of them is definitively above this limit, the maneuver will be forbidden. Parallel to the limitation of the safety envelope, the

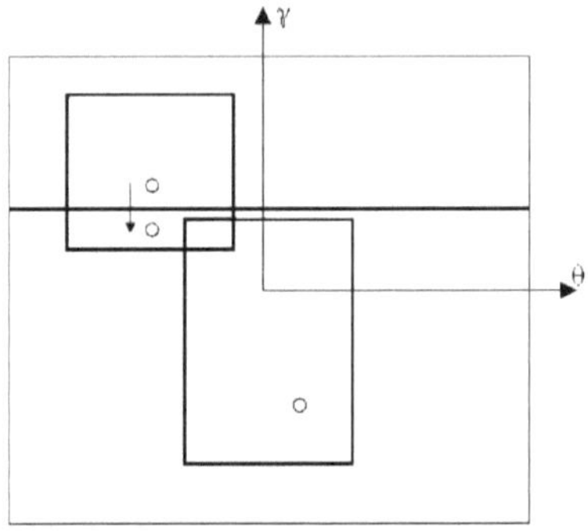

Fig. 3.1. Result of the preventive action on the reactive safety envelope

position of the optimum has to be checked too. If one optimum has a higher acceleration than allowed, its acceleration will be saturated to be under this limit with a safety gap. Finally, if the acceleration is not enough to create a static gap, the middle of the acceleration range will be used as optimum.

3.1.1 Environment categorization

The tactic agent presented in this section is specific to road portions without intersections like highway, speedway or some urban streets. As the intersections are not taken in the hypotheses, the agent has only to look at the vehicle flow on the lanes. Thus, the agent can monitor both speed and distance on the lane, and in addition it can control the lane changing. Moreover out of urban places, driving rules are almost respected. However, it is not necessary to adopt a probabilistic approach to look at other less possible scenarios. However, this probabilistic approach will be required in the city, where sudden scenarios can appear even if they are not allowed theoretically. By analyzing deeper in details the road accident statistics [122], the topology on such road portions comes into sight. Actually collision involving both vehicles changing their lane is really seldom appearing (about 1.7%). Therefore, the environment characterization can focus just on the current lane and the two adjacent lanes.

The analysis of these three lanes is performed by loading first the driving rules depending on the country and the weather conditions. After that the analysis will be looking at the road properties such as the curvature, the widths of the current lane and the markings. The width of the two other lanes is rarely measured directly, even if some projects such as [11] have shown positive results. So, if this information is not available, the width of the current lane will be duplicated to the other lanes. Normally, these widths are nearly the same, except the emergency stop lane, which is 2.75 m wide and not 3.25 m. In this uncertain case, the extreme limits of the safety envelope corresponding to the lane changing limits will be roughly approximated. This lack of accuracy is not dangerous, because this second-order information will be determined with a better accuracy as soon as the vehicle changes lane. Evenmore, the agent in charge of this computation can be faster and less precise than the other agents.

Following this, the avatars corresponding to the different road users will be posed on the pattern (figure 3.2(a)). On the six zones ahead and behind the vehicle, the closest road user's avatar will be stored to focus only on the short range environment. The platoon and its effect will be neglected here. Furthermore with the introduction of a C2C system, the pattern may be extended easily. When a vehicle is on two lanes at the same time (lane changing, overtaking), both lanes will be fully occupied by this vehicle. Next, any vehicle will be replicated as ghost on all the lanes on its right. This artifice will avoid having a vehicle passed on the right side: a minimal safety distance of 3 m will let the ego vehicle staying behind the other vehicle. Nevertheless, if another vehicle is occupying the lane, both distance control maneuvers will be performed. However, all these procedures will be automatically stopped when the situation becomes dangerous (a TTC lower than a second [77]). At the same time, lane changing by continuous marking and over-speed will also be allowed. Indeed, this voluntary neglect of the driving rules improves the possibilities of action for the driver in danger.

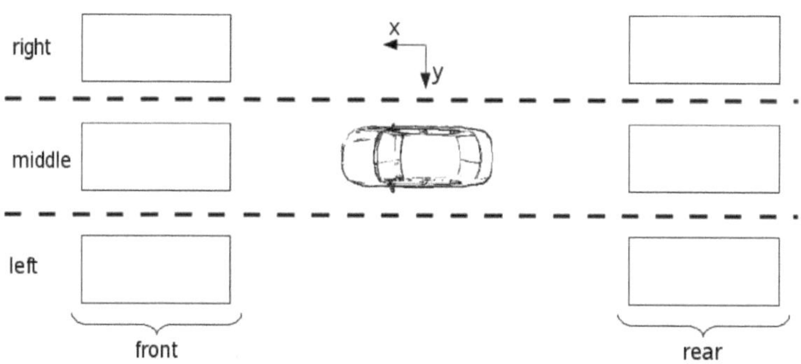

(a) Pattern integrated into the reactive part

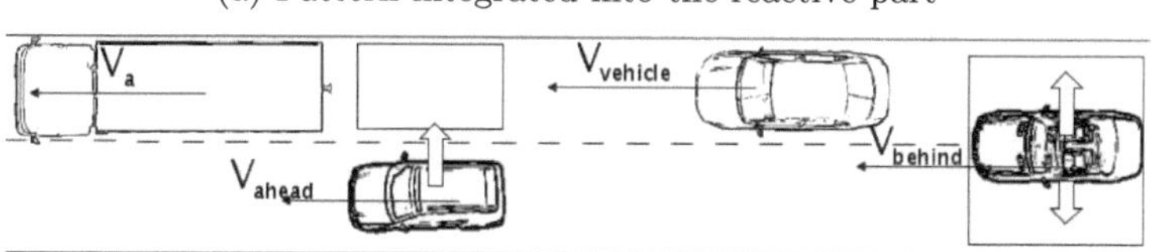

(b) Application of avatars on the environment pattern

Fig. 3.2. Example of environment characterization by use of avatars

3.1.2 Choice of the longitudinal and lateral controllers

Once the avatars matched on the environment pattern, the next step is to extract the maneuvers that can be performed. Reusable longitudinal and lateral controls can be split to fuse together one longitudinal controller and one lateral controller for each maneuver (distance control, lane changing etc.). In fact, the computation of the safety envelope has to be maneuvers based, and these maneuvers are lanes and objects depending. Thus, the computation of the safety envelope can use the information of the lanes to determine the extreme steering ranges, and this computation needs to use the avatars' information for the acceleration range.

Theoretically, only two agents are required for the steering ranges as in the figure 3.3(a). The first algorithm will define the maximal steering angles to stay in a lane (the current lane or any other lanes). The second algorithm will define the minimal angle to change the lane if there is another vehicle or not. In practice, four different agents will be integrated. Indeed, the optimization of the first agent requires the split between lane in the curve and lane out of the curve: it is easier to tend to go out of a curve rather than to go into a curve when the speed is not adequate for the road curvature.

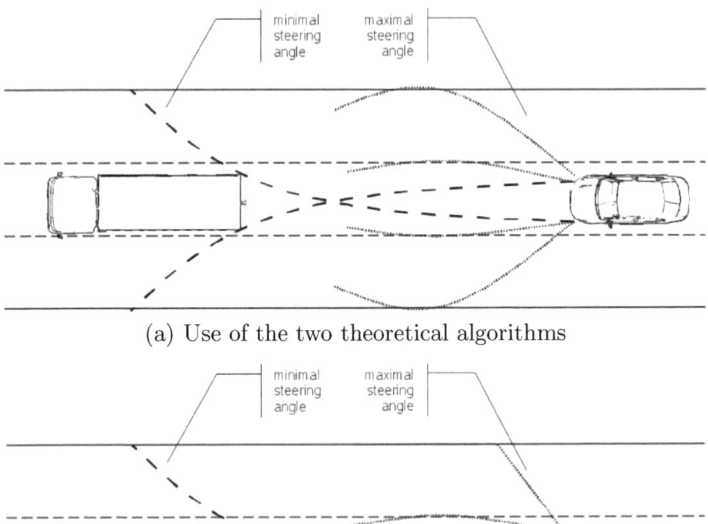

(a) Use of the two theoretical algorithms

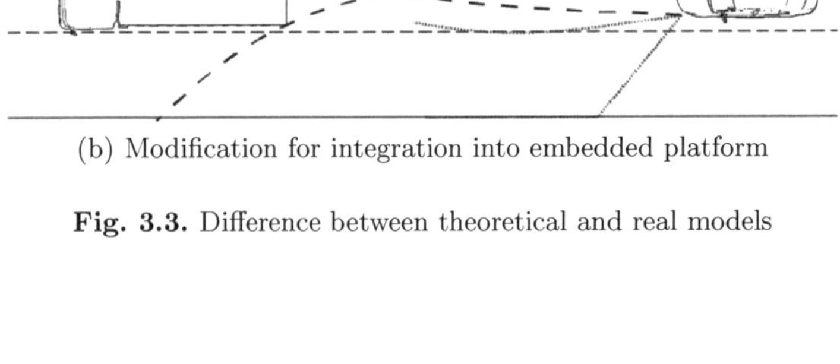

(b) Modification for integration into embedded platform

Fig. 3.3. Difference between theoretical and real models

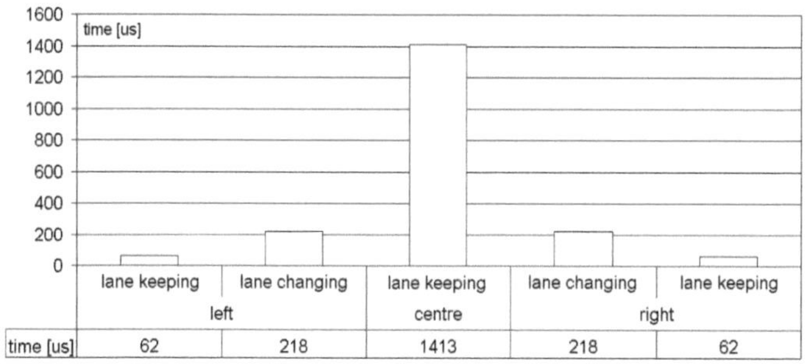

Fig. 3.4. Computation time of the different agents for lateral control on a Freescal MPC 5554

Furthermore the low accuracy of the limits position of the other lanes makes this high computation almost useless. So, an alternative agent will perform this computation with lower resources and also lower non-vital precision (Fig.3.4).

3.2 Longitudinal controllers

The longitudinal controllers which are required are split into two categories. For each maneuver, two extreme (emergency brake) and an optimum (e.g. the ACC) have to be determined. The relevant maneuvers requiring this information are the following. The distance to the vehicle ahead will define a maximal acceleration. When a vehicle behind the ego-vehicle exists, a minimal acceleration (actually a deceleration) has to be determined too. Last but not least, no vehicle should be passed on the right as long as the situation is not dangerous. Thus a maximal acceleration needs to be defined to have the ego-vehicle staying a few meters behind the other vehicle on its left.

3.2.1 Safety acceleration for the front direction

Up to now, collision mitigation systems are waiting that accidents cannot be avoidable anymore to generate a full braking. However, the next generation of emergency brake will move to collision avoidance by limiting the acceleration depending on the distance and both speeds.

Let us take a distance d_0 to the vehicle ahead. The ego speed will be named v_0 and the speed of the vehicle ahead will be named v_a. For electrical wedge brake, it is possible to simplify the dynamical model to a first-order transfer function with the following equation for the acceleration and depicted in figure 3.5(a):

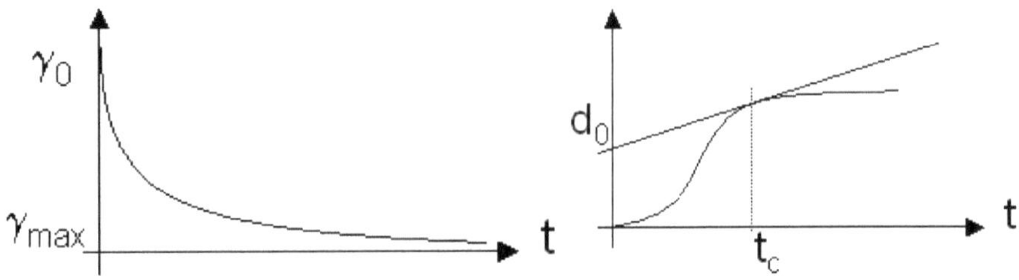

(a) Acceleration model over the time (b) Distance evolution over the time

Fig. 3.5. Integration of a first-order model for the acceleration to approximate the distance over the time

$$\gamma(t) = (\gamma_{max} - \gamma_0) \exp -\frac{t}{\tau} \tag{3.1}$$

Avoiding a collision at the very last moment means that the vehicle is allowed only to touch the other vehicle. In practice a safety distance d_{min} is, of course, taken into account. Mathematically spoken, the function describing the distance over the time has to have null global minimum as described in the figure 3.5(b). Let us name this critical moment t_c where the vehicle is supposed to touch the other one.

First of all, the global minimum has to be determined to find the position of t_c. Thus the zero crossing of the derivation of the function distance has to be found.

$$v_a = v_0 - (\gamma_{max} - \gamma_0)\tau \exp -\frac{t}{\tau} \tag{3.2}$$

$$\begin{cases} t_c = \tau \cdot ln\frac{-\tau(\gamma_{max}-\gamma_0)}{v_a - v_0} \\ if \frac{\tau(\gamma_{max}-\gamma_0)}{v_a - v_0} < 0 \end{cases} \tag{3.3}$$

Next the distance at this time t_c has to be minimal:

$$d_0 + v_a \cdot t + d_{min} = v_0 \cdot t + (\gamma_{max} - \gamma_0)\tau^2 \exp -\frac{t}{\tau} \tag{3.4}$$

Unfortunately this equation cannot be solved mathematically. Thus a look-up table is used in practice to solve this equation in real time.

3.2.2 Distance control for the front direction

The maximal speed (v_{max}) is defined when the minimal distance will be reached anyway because of a hard braking. In this case the distance will be lowered by the distance (Δ_{min}) at a time (t).

$$\Delta = d_0 + (v_0 - v_{max})t - \frac{\gamma_{max}}{2}t^2$$

$$v_{max} = v_0 + \sqrt{8 \cdot \gamma_{max}(\Delta_{min} - d_0)} \tag{3.5}$$

The vehicle gets more freedom to move longitudinally than laterally because of the small width of the lane. Actually no comfort range for the steering angle can be defined directly because the safety range is too small. The optimum is simply defined as the middle of the steering range in order to always get the vehicle away from lanes' border.

However, a comfort range for the longitudinal control can exist if the density on the road is not important. Here the comfort range is defined between a maximal acceleration which describes the realization of the normal safety distance to vehicle ahead, and a minimal acceleration which describes a hard braking. Comfort range defined in terms of lateral and longitudinal max accelerations (limits different because two different muscles used). Standard values can be defined as $| \gamma_L | \leq 3m/s^2$ and $| \gamma_l | \leq 1m/s^2$ by [92] and extended over the time.

3.3 Lateral controllers

For the lateral controllers, the same kind of maneuver definition is also used. On top of the steering optimums for the lane keeping and for the lane changing (on the right and on the left), steering ranges are also required. Here only these range definitions are described since the concepts are new. Information about the determination of the optimum can be found by looking at the algorithms for autonomous vehicles [31].

3.3.1 Safety range for the lane keeping

The mathematical model is built upon eight parameters, seven input parameters (curve radius R, road width d, relative position to mid-lane y_0, initial velocity v_0, acceleration γ, yaw orientation δ and steering dynamic SV) and one output parameter (θ) which create the safety envelope (see geometrical model fig. 3.6).

One elementary equation, $h(t,\theta)$ stands for the minimal distance between vehicle and road border, and defines the mathematical background of all three models below. $-d$ is for the left road border and $+d$ for the right one.

$$h(t,\theta) = \sqrt{(x(t,\theta) + x_0)^2 + (y(t,\theta) + y_0)^2} - (R \pm d) \tag{3.6}$$

Two unknowns (t,θ) lead to two conditions in order to fix the point where $h(t,\theta) = 0$ and $\dot{h}(t,\theta) = 0$:

$$\begin{cases} \dot{h}(t,\theta) = 0 \rightarrow t_{min} = ... \\ h(t_{min},\theta) = 0 \rightarrow \theta = ... \end{cases} \tag{3.7}$$

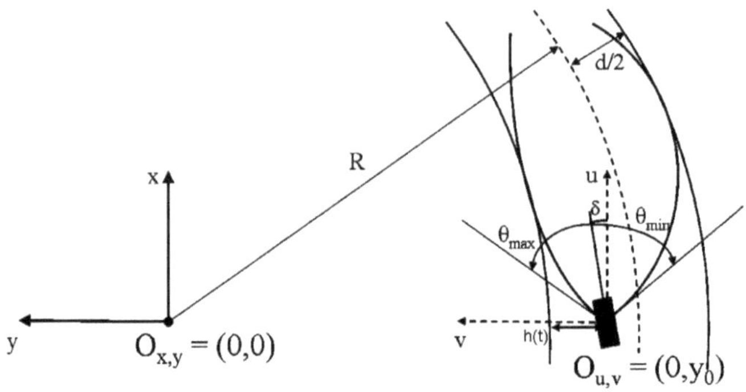

Fig. 3.6. Geometrical model for the lane keeping controller

The main objective here is to define t and θ so that only one collision point is possible with the road border. First the vehicle speed has to be defined as below:

$$\begin{cases} v_x(t,\theta) = (v_0 + \gamma \cdot t) \cdot \cos(\theta + \delta - SV \cdot t) \\ v_y(t,\theta) = (v_0 + \gamma \cdot t) \cdot \sin(\theta + \delta - SV \cdot t) \end{cases} \tag{3.8}$$

By integration over the time, the vehicle coordinates $(x(t,\theta), y(t,\theta))$ are defined. To prevent unexpected physical elements, such as variations of friction coefficient μ, a security distance $d_{security} = 10\ [cm]$ is introduced, in which the vehicle can theoretically not penetrate.

Different approaches

Three methods have been used to solve equation (3.8). They will be presented now and comparison will be done to choose the one that will be integrated within the agent.

Iterative model

Its main advantage is its high accuracy. This model is mainly used to serve as exact referential model. The iteration step is defined by the minimal executable steering angle.

Approximate model

The idea is to create an online and direct method for a general usable algorithm (different vehicle environmental circumstances). Therefore, symbolic expressions containing the eight mentioned parameters are the fundaments of this model. Main compromise is due to mathematical complexity and high accuracy against computation power. The latter depends strongly on hardware constraints. Therefore, approximations are adapting the model and reduce the mathematical complexity. Taylor series are used in the vehicle coordinate (x, y) expressions for the trigonometric functions and the square root estimations. Taylor series are limited to degree 3. Thus two solutions will be mathematically possible: $t_{min,1}$ and $t_{min,2}$.

The square root approximation based on Taylor is responsible for the most important error on the time expression. Both expressions, $t_{min,1}$ and $t_{min,2}$ show an error lower than 30% (fig. 3.7 left) for $\theta \to 1\,[rad]$. This error induces only small variations for the steering angle.

The time scale of t_{min1} is negative because t_{min2} is the positive solution in this case. The coordinates are transformed to polynomial expressions in order to facilitate the control of the equation complexity and of the development.

Look-up table model

The creation of the look-up table is based on the same algorithm used for the iterative model. The only difference lies in the parameter range which is limited to the most significant values. These values are defined with the help of the error analysis in order to acquire best ratio accuracy against memory cost of the look-up table. The size of this table is limited by hardware, more precisely by the allocated memory space (ROM). The compromise lies in accuracy against matrix size. The bigger the look-up table is, the more accurate the estimation can be.

The seven basic parameters ranges have been defined on the error analysis and are following:

- $\gamma_0 \in \{-5 \quad 0 \quad +5\}\ [m/s^2]$
- $v_0 \in \{30 \quad 55 \quad 80\}\ [km/h]$
- $\delta_0 \in \{-10 \quad +10\}$ [grad]

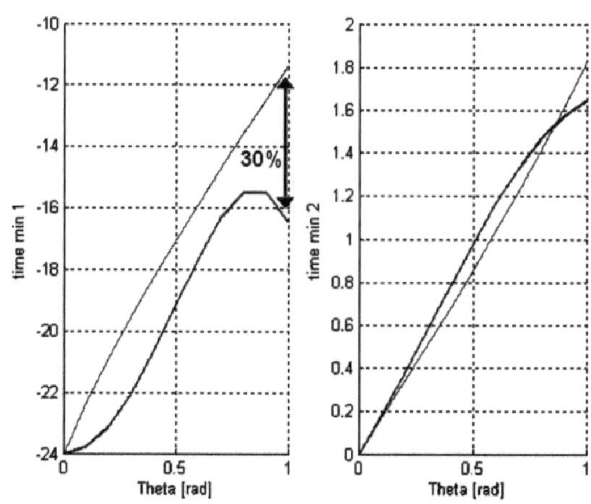

Fig. 3.7. Comparison of $t_{min1,2}$: t_{min} under ratio form and t_{min} with Taylor approximation for $Q(\theta)$

- $SV \in \{-5 \quad -15\}$ [grad/s]
- $R \in \{10 \quad 15 \quad 20 \quad 30 \quad 50 \quad 100 \quad 500 \quad 1000 \quad 10000\}$ $[m]$
- $y_0 \in \{-1 \quad 0 \quad 1\}$ $[m]$
- $d \in \{2.75 \quad 3.75 \quad 7.5\}$ $[m]$

The acceleration/deceleration of personal cars does not exceed 5 $[m/s^2]$. The algorithm will not be used here beneath 30 $[km/h]$. 80 $[km/h]$ is a maximal value for urban environments. The vehicles yaw orientation for $v_0 \geq 30$ $[km/h]$ is rarely higher than the given value. The wheel angle dynamic $SV = 10$ [grad/s] corresponds also to a maximal values. The curve radius range comports the most values and they are concentrated between 10 and 100 m. The steering angle is especially sensitive to R. The small radius approximates close curves whereas the high radius approximates almost straight road segments. The road width value corresponds to common freeway and highway values. 7.5 $[m]$ is used to approximate double lane roads. Extensions for all parameter ranges can be added later on but required memory space and execution time will increase in consequence.

Linear interpolation is used in order to approximate the steering angle depending on the seven dimensions within the look-up table. This method is the fastest and gives good approximation accuracy because single interpolation (see section 3.1 of this chapter) is almost linear for the most parameters. And time-error comparison between linear and cubic interpolation results in:

The steering angle varies in a quasi-linear manner in function of each basic parameter. The values in the parameter ranges are closed from each other and do not permit curvy segments between two successive values. The calculated mean error of linear interpolation (table 3.1) of 2.5% is acceptable and will be corrected by optimization of $d_{security}$. However, the most important criterion is the computation time which shows that linear interpolation is the fastest method.

Model validation and comparison

The goal is to evaluate in the first step the error due to single interpolation which means that only one parameter out of the seven basic parameters needs to be approximated by interpolation. This method will show the weight of dependency of the error for each basic parameter. A second step consists in making multiple interpolations and to synthesize a maximal error, a worst case. This method shall show a theoretical limit of the model. For all comparisons and error analysis, only collisions with the left road boarder are considered.

Table 3.1. Comparison of computation times

Method	Normalized time [-]	Error mean value [%]
Linear	1	2.5
Cubic	4.3	0.92

Error analysis between the look-up table and the iterative model

First only one parameter varies; the error analysis focuses on $\theta(p)$ where p is one of the seven basic parameters (γ_0, v_0, δ, SV, R, y_0, d). This part serves to illustrate the weighting of each parameter on the steering angle θ.

The steering dynamic SV graph (figure 3.8(a)) shows that the values from the interpolation are rarely exact, the error graph is disquiet but regular. The minimal iteration step can be observed by means of the graph with its stair form. The red line connecting the three predefined values of SV build a straight segment and two values are sufficient for correct results.

The distribution of the first predefined values for the curve radius 10, 30, 50, 100, 500, 1000 and 10'000 m for the look-up table are concentrated around the small values, limited to $10 \leq R \leq 100$ $[m]$); the big values serve principally for the approximation of straight road segments. The error analysis is therefore more representative. The most significant error is found at around $R = 20\,[m]$; it rises up to 8%. By adding 15 [m] and 20 [m], the error is reduced to 2%.

With the help of the previous paragraph an hypothetical worst case is synthesized. The exact value of the steering angle θ is calculated by the iterative model. This section will then analyze the error of the worst case. Imaginary worst case is synthesized out of the single interpolation. This specific situation leads to an error of 11.5%. Because the basic parameters are linked together, this value serves to give an order of magnitude of the error and permits to dimension the value of $d_{security}$.

Error analysis between the approximate model and the iterative model

The same mean values for the seven basic parameters are used for simple interpolation. The biggest error is caused by $\theta(\gamma)$ where the maximal error interval will be defined.

The error in the hypothetical worst case of multiple interpolations is 8.5% and is acceptable. This value has not to be considered as maximal possible error because the influence of multiple interpolation on all values cannot be analyzed (parameters are coupled).

Conclusion

The error analysis of the interpolations shows that the look-up table model error is comprised in the interval -1.5 and $+2\%$ and that the approximate model error between -5 and $+5\%$. Furthermore, the interval of the approximate model is characterized by high errors due to the parameter R. The reached accuracy of the approximate model is satisfying but the main disadvantage resides in the complexity of the symbolic expressions and the high

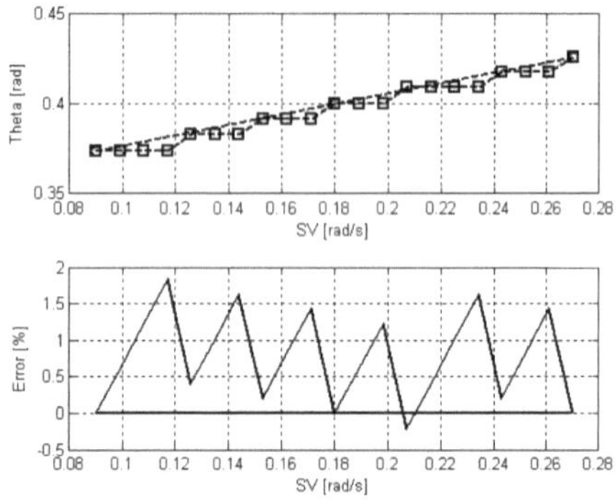

(a) Error analysis with single linear interpolation, $\theta(SV)$. Doted: look-up table with linear interpolation; doted: iterative model

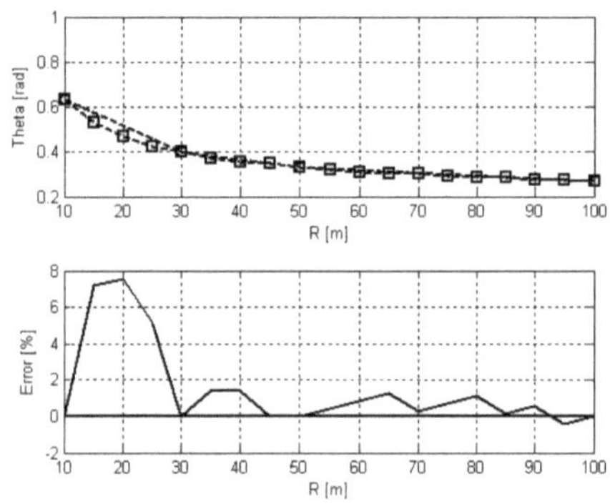

(b) Error analysis with single linear interpolation, $\theta(R)$. Doted: look-up table with linear interpolation; doted: iterative model

Fig. 3.8. Analysis of the error due to the approximation and linearization

computation time. In order to downsize both points, either the approximation order has to be decreased or the number of basic parameters has to be decreased. The main characteristic of the look-up table is that the accuracy level is proportional to its size. For this reason, the size of the look-up table has to be put to its maximum to obtain finally a maximal accuracy. The two worst cases studied are not representative because the error of single interpolation was biased on mean values. In addition to that, the basic parameters are coupled: by changing more than one parameter, the result cannot be predicted (by comparing with single interpolation). The error percentages of 11.5 and 8.5% give anyway an order of magnitude of what maximal error could be (Fig. 3.9).

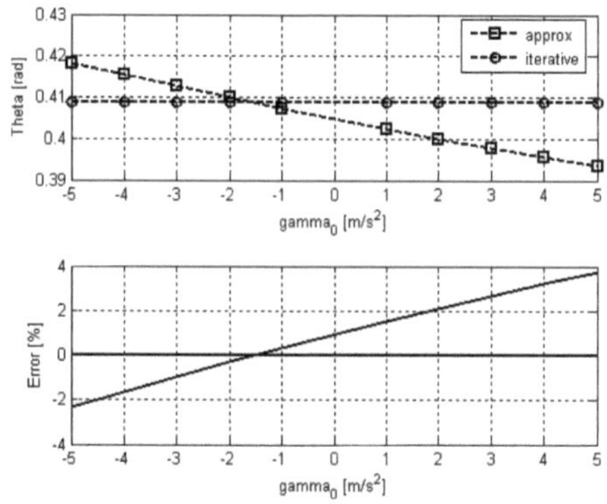

Fig. 3.9. Error analysis with single linear interpolation, $\theta(\gamma)$

3.3.2 Extreme lane keeping assistant for other lanes

The alternative method will be used to estimate roughly the extreme lane changing. Lane changing maneuvers have been measured with different speeds and drivers. General lower bounds and mean values have been estimated depending only on the speed as depicted in figure 3.10. The first value describes indirectly the abrupt lane changing to avoid collision with another vehicle, and the second time corresponds to a standard comfort lane changing maneuver. But over this time, the steering angle did not stay constant, to help the angle determination with a simple trigonometric calculation. To solve this problem, additional visual information have been used: for the lower-bound case, the maximal steering angle is corresponding to an alignment of the vehicle with three lanes after. Thus, all in all, it is possible to calculate directly the maximal steering angle θ_{max} by using the lane changing time $\mathcal{T}$ and the lane width w times 3 (see equation 3.9).

$$\theta_{max} = 3 \cdot \mathcal{T}(v, w) \tag{3.9}$$

3.3.3 Safety distance for the lane changing

When there exists a risk of collision with another vehicle (ahead or behind the ego vehicle), a minimal steering angle can define a lane changing to avoid this collision. Let us use the geometrical model depicted in the figure below.

The time to collision (t) can be defined with the driven distance (l) by using at the following equation:

$$\frac{\gamma}{2}t^2 + (V_o - V_a)t - d = 0$$

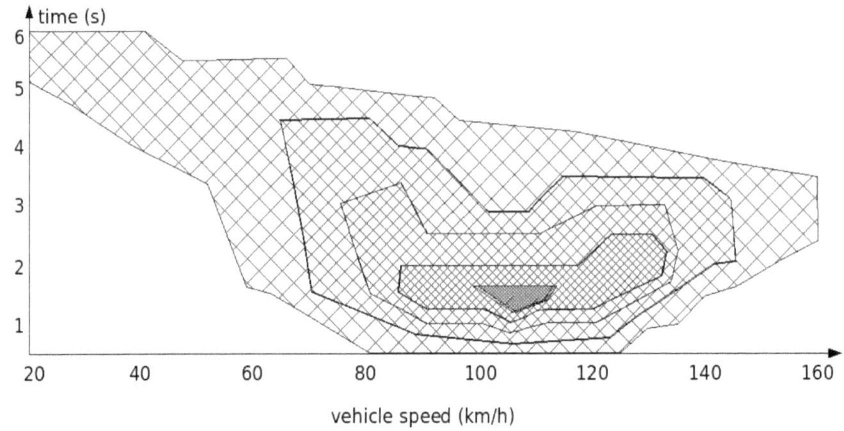

Fig. 3.10. Relation between speed and realization time for overtaking maneuvers

$$\begin{cases} t = \frac{2(V_a - V_o) + 2\sqrt{V_a^2 + V_o^2 - 2 \cdot V_a \cdot V_o + 2 \cdot \gamma \cdot d}}{2 \cdot \gamma} \\ l = V_o \cdot t + \frac{\gamma}{2} \cdot t^2 \end{cases} \tag{3.10}$$

If the tire dynamics and the steering rate are neglected, the vehicle will perform a circle when it gets a constant steering angle. This circle will go through the current vehicle position with a normal tangent. It will also go through the point (F), which is at the same distance as the collision point (I) but on the other lane (Fig. 3.11).

Let us declare the following intermediate variables:

$$R^\star = R + \frac{w}{2} - \delta \ , \ \theta = \frac{l}{R^\star}$$

Then the contact point (I) can be defined on the (x,y) plane. After that, the point (F) can be also determined on the same plane.

$$I \begin{cases} R^\star sin(\theta) \\ R^\star(1 - cos(\theta) \end{cases} , F \begin{cases} (R - \frac{w}{2}) sin(\theta) \\ R + \frac{w}{2} - \delta - (R - \frac{w}{2}) cos(\theta) \end{cases} \tag{3.11}$$

Next, the equation of this circle can be used to define its radius (R_c).

$$(x - 0)^2 + (y - R_c)^2 = R_c^2 \ , \ R_c = \frac{F_x^2 + F_y^2}{2 \cdot F_y} \tag{3.12}$$

Fig. 3.11. Model for the distance control

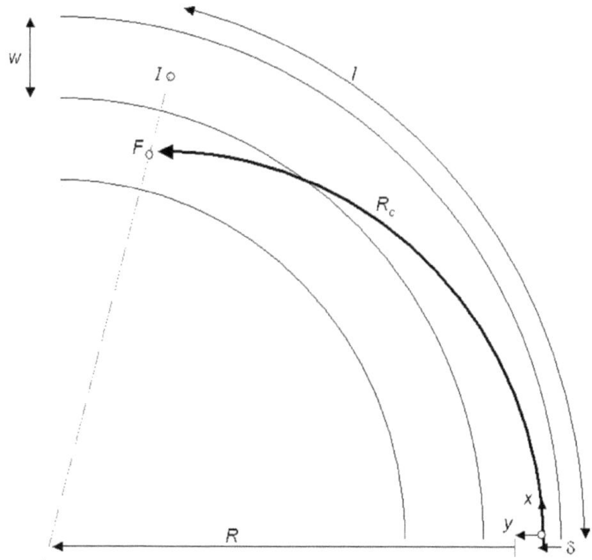

Fig. 3.12. Model for the lane changing

Finally, the Ackerman model (see fig. 4.7, p. 67) will be used to determine the steering angle (δ).

$$tan(\delta) = \frac{L}{R_c + w/2} \tag{3.13}$$

If there is no other vehicle, the minimal steering angle would be defined by touching the other lane. However, such definition is not very helpful. For example, the minimal steering angle for lane changing on a straight forward road is null as the lane can be touched at the horizon without steering intervention. That is why the human model defined previously in section 3.3.2 of this chapter will be reused here to define a general time for the lane changing maneuver (fig. 3.12).

3.4 Anticipatory action to prevent inappropriate speed

In front of the vehicle, many things can influence the vehicle speed and may require the driver to adapt the vehicle speed in advance: stop shields, traffic lights, speed limitation, road curvature etc. All these elements can be taken into account on a speed graph ahead of the vehicle. First the influence of the road curvature will be taken into consideration to limit physically the speed. Then the other local speed limitations will be added on this graph. At the end, the vehicle dynamics will be used to define the maximal acceleration allowed to stay in the speed envelope.

The maximal speed defined here is only validated when the driver stays parallel to the marks in a curve. Nevertheless it will be possible to get a higher speed if the driver realizes a higher radius by translating the vehicle from the outer border to the inner border.

3.4.1 Computation of the maximal safe speed

Rolling arises if there is a torque around the longitudinal axis of the vehicle. The reason for this torque can be lateral forces or different loads of the left and right side of the vehicle. In figure 3.13 the rotation of the vehicle body around the rolling center (CoR) is sketched. To be able to integrate the rolling angle (κ) into the model the vehicle rolling has to be described dynamically. The suspension ($c_{l,r}$) and damper ($k_{l,r}$) elements as well as possible stabilizers (c_S) mounted on the axles produce forces and torque which work contrary to the rolling. If the height of the center of gravity h_{CoG} and the height of the rolling center h_{CoR} are different, the rolling angle κ will be related to the lateral acceleration γ_l:

$$J_\kappa \cdot \ddot{\kappa} + k_w \cdot \dot{\kappa} + (c_w + c_s)\kappa + M_R \cdot sign(\dot{\kappa}) = (h_{CoG} - h_{CoR}) \cdot m \cdot \gamma_l$$

Because of the use of the bicycle model, the parameters k_w, c_w and c_s represent the average values of all damper, suspension and stabilizer elements. The parameter J_κ describes the inertia of the vehicle body around its longitudinal axle, while m is its body mass. With M_R, possible suspension friction

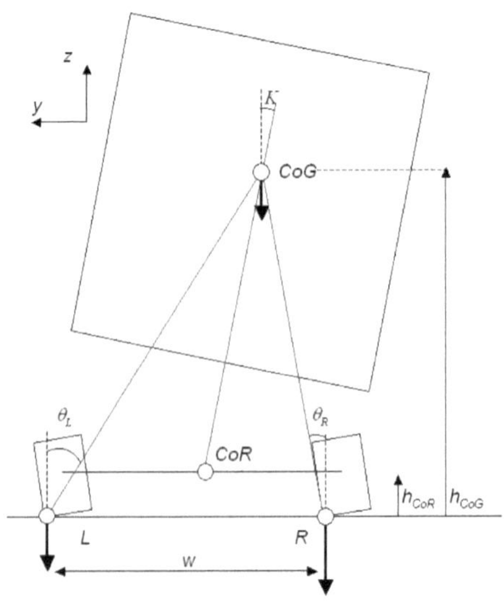

Fig. 3.13. Torques creating the risk of rolling

can be respected. Further on, here the vertical elasticity of the tires can be built in by a skilful reduction of c_w.

If the movement of the vehicle is approximated to be constant, the lateral acceleration will be only the centrifugal force and κ is constant:

$$\kappa = m\frac{h_{CoG} - h_{CoR}}{c_w + c_s}\frac{v^2}{R} \tag{3.14}$$

The gravity force is defined at the center of gravity, and the vehicle has two contact points per axle with the ground on L and R. Therefore the forces on the ground have their axles crossing on the center of gravity and the vectors are forming a triangle with the gravity force. Thus by defining the position of the center of gravity these two angles are known:

$$CoG \begin{cases} h_{CoR} + \frac{h_{CoG}-h_{CoR}}{cos(\kappa)} \\ (h_{CoG} - h_{CoR})tan(\kappa) \\ \end{cases}$$

$$\begin{cases} \theta_R = \dfrac{h_{CoR}+\frac{h_{CoG}-h_{CoR}}{cos(\kappa)}+\frac{w}{2}}{(h_{CoG}-h_{CoR})\tan(\kappa)} \\ \theta_L = \dfrac{h_{CoR}+\frac{h_{CoG}-h_{CoR}}{cos(\kappa)}-\frac{w}{2}}{(h_{CoG}-h_{CoR})\tan(\kappa)} \end{cases} \tag{3.15}$$

By using the sum of the forces $m\mathbf{g}$, $\mathbf{F}_L$ and $\mathbf{F}_R$ equal to zero, the following equations are obtained

$$\begin{cases} F_L \cdot cos(\theta_L) + F_R \cdot cos(\theta_R) = m \cdot g \\ F_L \cdot sin(\theta_L) + F_R \cdot sin(\theta_R) = 0 \end{cases} \tag{3.16}$$

The vehicle will start rolling to the right when F_L is null. That means that θ_R is null and

$$h_{CoR} + \frac{h_{CoG} + h_{CoR}}{cos(\kappa)} + \frac{w}{2} = 0$$

By reusing equation 3.14 and extracting κ from the equation ahead, the maximal speed can be defined as follows:

$$v = \sqrt{R\frac{acos(-\frac{h_{cog}-h_{cor}}{h_{cor}-w/2})}{m \cdot (h_{cog} - h_{cor})}(c_w + c_s)} \tag{3.17}$$

This plan model is based on the hypothesis of the influence between front and rear axles and can be neglected. However, the model needs at least for trucks with semi-trailer to be improved by use of a multi-body approach. In practice the rolling effect from the trailer cannot be neglected at all for the truck for small curvature. Thus, the mass body has to be split on both axles as shown in the figure below. After that the stiffness of the chassis can be taken into account to define a resistive term (Fig. 3.14).

$$m^{\star}_{front} = m\frac{l_{rear}}{L}, m^{\star}_{rear} = m\frac{l_{front}}{L}$$

Speed graph correction with the maximal braking capacity

The curve of the maximal speeds must be corrected by taking into account the dynamical braking capacity of the vehicle. For each part of the curve in front of the local speed minima, it is necessary to limit linearly the speed depending on the maximal feasible deceleration and maximal comfortable deceleration as in figure 3.15. Otherwise it would be possible to get a safe speed at the current time but the driver will be in danger later because the vehicle will not be able to brake enough to get an adapted speed later.

Determination of the maximal acceleration

The maximal acceleration can be defined now by linking the information coming on one side from the current speed, and on the other side from the speed graph (current maximal speed, distance to the local minimum and the corresponding speed). The additional information needed is the braking dynamic like shown in figure 4.3 (in Chapter II.4). Here a static braking rate can be

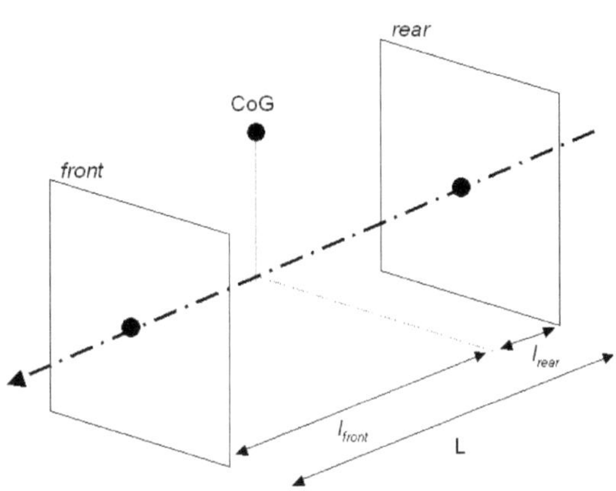

Fig. 3.14. Multi-layer model for the integration of the chassis stiffness

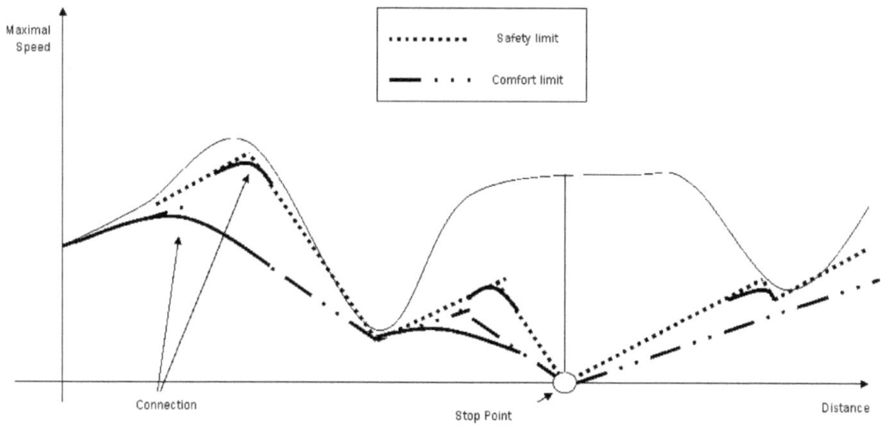

Fig. 3.15. Graph of maximal safe speed

taken as assumption and the latency time (due, for example, to the movement of brake pads) does not have be taken into account because it can be pre-compensated after that by sending earlier the information of hard braking. Furthermore for extreme cases this model will not be accurate enough and the direct use of the transfer function of the braking is required. This function can be resolved to provide the speed over the time and thus the speed over the distance.

3.4.2 Extension to multiple paths

This application determines the maximal acceleration as soon as the road topology is defined. However, doubts are still possible as soon as the vehicle comes close to a crossing section. Indeed at such moments, several routes with different maximal acceleration are possible. An add-on has to estimate which route will be chosen by the driver, even if the information from the navigation is not provided. This recurrent problem at the exits on highway is unfortunately fundamental because, the speed often needs to be adapted before the vehicle changes the lane to go on the dedicated lane.

Parameters for the estimation of the direction

Three different routes will be extracted from the road tree model extracted from the navigation platform: the main route and the two closest crossing sections as they correspond to most situations. The algorithm can be extended easily to many more connections by modifying the rules for the Bayesian estimation.

Most of the drivers will prepare the exit of a highway about 1 km before by positioning the vehicle on the right lane. Therefore a double lane changing in the direction of the exit near a crossing section can be seen as sign for road changing. Thus, the interesting parameters used here are the blinkers, the lateral speed and the lateral distance to the changing lane.

Furthermore, most of the drivers also slow down before a junction. Indeed as shown in figure 3.16(a) and 3.16(b), the vehicle speed for several tests drivers (performed only in Germany) is defined depending on the distance. For the speedway two main deceleration models can be extracted depending on the exit topology: can the driver brakes later or not. However, the speed will not been taken into account because a rule can only be validated on a highway, where homogeneity can be seen.

With the help of the influence of these data, independent probabilities can be computed and their product leads to an estimation of the route choice. Actually the parameters lateral distance and lateral speed are not completely independent as their combination generates the maneuver. To simplify the computation, they will be set independent as in the equation below, which is used for each route (i).

$$\mathcal{P}_i = \mathcal{P}_{indicators} \cdot \mathcal{P}_{speed,\ lateral} \cdot \mathcal{P}_{distance,\ lateral} \tag{3.18}$$

The route with the highest probability will be defined as the route chosen by the driver. However, this binary choice will provide discontinuities for the maximal acceleration. Therefore the computation of this maximal acceleration will base on the weighting of the maximal acceleration of each route (main, left, right) with the corresponding probabilities (Fig. 3.17):

$$\gamma_{max} = \frac{1}{\sum_i \mathcal{P}_i} (\mathcal{P}_{m.} \cdot \gamma_{m.} + \mathcal{P}_{l.} \cdot \gamma_{l.} + \mathcal{P}_{r.} \cdot \gamma_{r.}) \tag{3.19}$$

Indicators

Indicators are a good advice as they express directly the driver's desire, but they can be omitted time to time. Therefore the algorithm must perform a good estimation even if the driver forgets to switch them on. If both are used together as warnings, the probabilities of the different possible routes will be set equal as the choice cannot be determined under this condition. Furthermore the absence of indicators is also an advice when the vehicle comes closer to the crossing. Therefore the use of the indicators needs to be correlated with the distance to the crossing and the duration of their use. Indeed an indicator switched on since several seconds will loose continuously the importance of its advice as the driver may have forgotten it. Figure 3.18(a) describes the corresponding truth table, whose parameters are depending either from the time $\mathcal{P}(t)$ (if one indicator is currently used) or from the distance to the crossing $\mathcal{P}(d)$ (if no indicator is used).

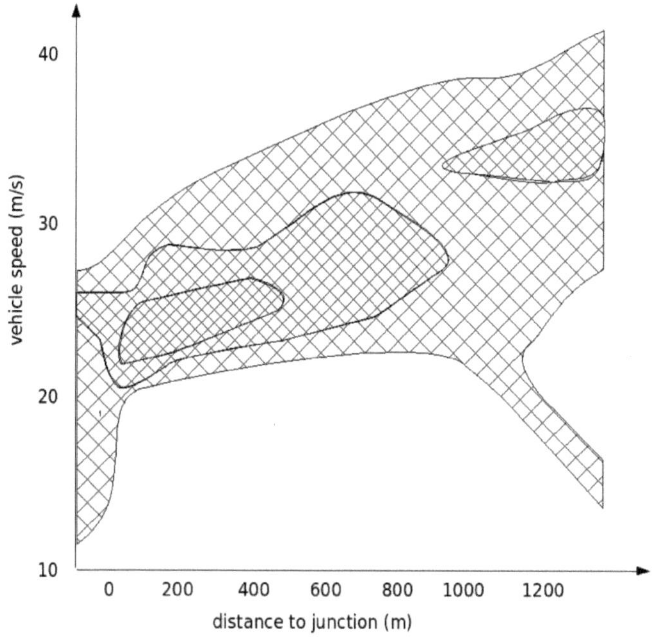

(a) Freeway

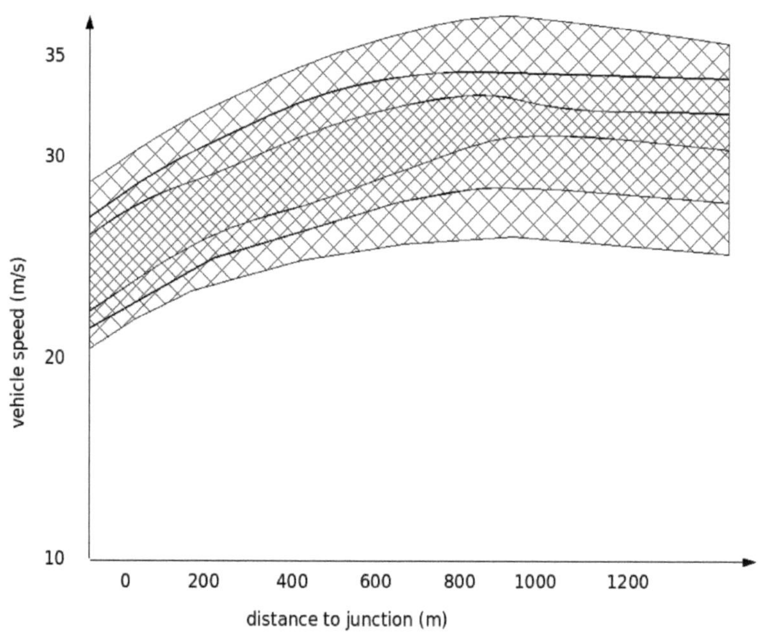

(b) Highway

Fig. 3.16. Speeds near a junction

Indicators		Route		
Left	Right	Left	Main	Right
0	0	$\frac{1-\mathcal{P}(d)}{2}$	$\mathcal{P}(d)$	$\frac{1-\mathcal{P}(d)}{2}$
0	1	$\frac{1-\mathcal{P}(t)}{3}$	$2\frac{1-\mathcal{P}(t)}{3}$	$\mathcal{P}(t)$
1	0	$\mathcal{P}(t)$	$2\frac{1-\mathcal{P}(t)}{3}$	$\frac{1-\mathcal{P}(t)}{3}$
1	1	0.33	0.33	0.33

Fig. 3.17. Properties of the routes depending on the indicators

Lateral speed

The lateral speed of the vehicle describes roughly the direction where the driver wants to go. Therefore a high lateral speed describes with an important probability the intention to change the route. This lateral speed ($v_{lat.}$) is measured from the vehicle speed (v) and the steering angle (δ) as in the following equation 3.20:

$$v_{lat.} = v \cdot sin(\delta) \tag{3.20}$$

The probability laws for the three routes are shown in figure 3.19(a). A probability of 99% can be set for a lateral speed higher than 3 $m.s^{-1}$ or -3 $m.s^{-1}$. This threshold has been found empirically after measurement with test drivers and it is corresponding to an important lateral deviation, which must be performed intentionally.

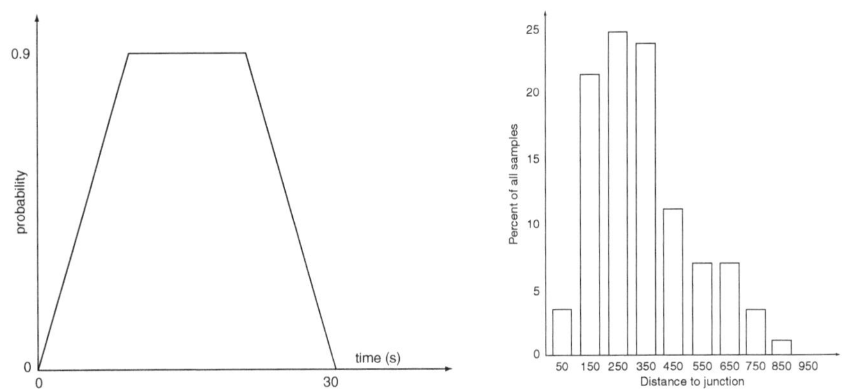

(a) Confidence of an indicator over the time

(b) Vehicle position at blink start

Fig. 3.18. Use of the indicators for the estimation of the driver's intention

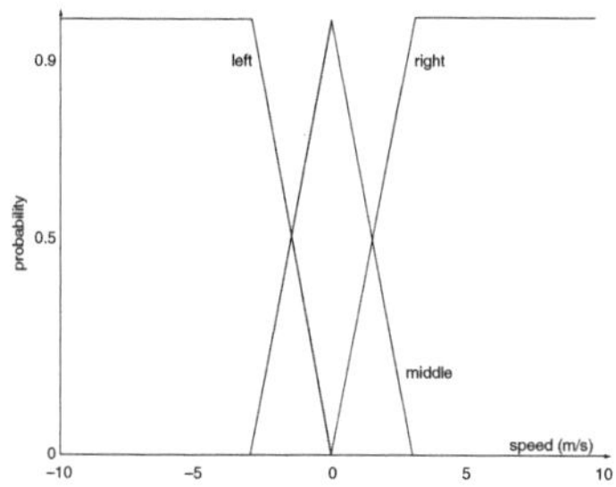

(a) Probability law for the lateral speed

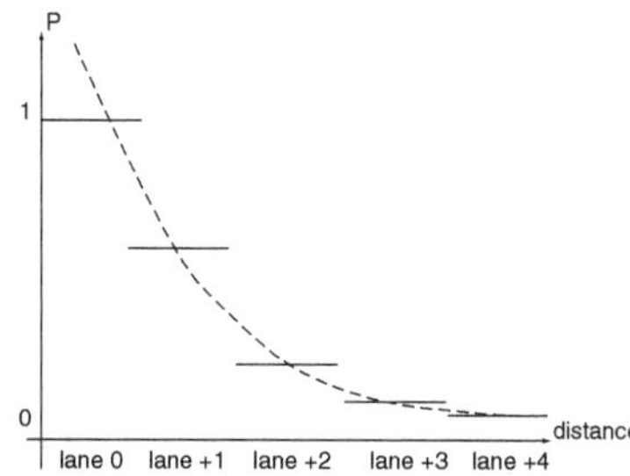

(b) Probability law for the lateral position

Fig. 3.19. Use of the lateral motion of the vehicle for the estimation of the driver's intention

Lateral distance

If there are several lanes on the road as on highways, the lateral position of the vehicle relative to the exit will be another parameter. Unfortunately this one is difficult to obtain without camera monitoring but it is very helpful. The probability law is inversely proportional to the lateral distance (Fig. 3.19).

Part IV

Adaptive cooperation

1

Methodology of a fault-tolerant adaptive cooperation

Abstract. This first chapter introduces the concept of adaptive cooperation and the required functionalities within the vehicle. After that, the safety aspects of this introduction is described to show which algorithms have to be used here.

1.1 Drawbacks of current emergency brake

The first system introduced in the market, which can intervene on the driver's command at the command level, has worked only on the longitudinal dynamics by means of the emergency braking. This choice is based on two facts. First the possible latency time for a braking is longer than for a steering intervention and, thus, a lower accuracy is required. Moreover an access to the braking actuators is already provided through the stability control (ESC) or with a braking booster. At the opposite a steering intervention requires a new steering actuator that can accept an external correction demand, the lateral vehicle positioning has to be very accurate and the intervention controller has to be very accurate and fast enough at the same time.

The initiation of an emergency braking has already been explained in a previous section (Chapter III.3, Section 3.2.1), using the distance and the relative speed with a vehicle in front. In the first versions of this device, the road–tire friction coefficient μ was not taken into account because it was not possible at this time to get a cost-effective and reliable sensing device. However, the next step will be its integration anyhow in order to improve the distance control (ACC) and above all this emergency brake. As this emergency braking is a safety device, it has been integrated right now alone as it must not be related to any non-safety function, whose reliability is lower.

The safety envelope is here limited to the maximal acceleration for the current lane. If the speed of the vehicle in front is high enough, the emergency braking will be performed at a very short distance and, thus, the driver really

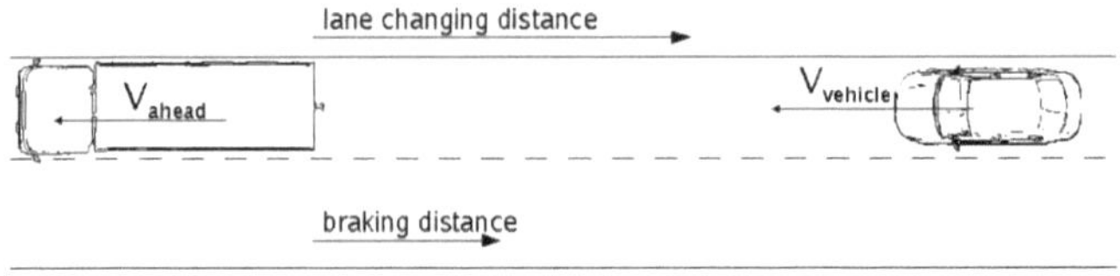

(a) Stopping distance and lane changing at high speed

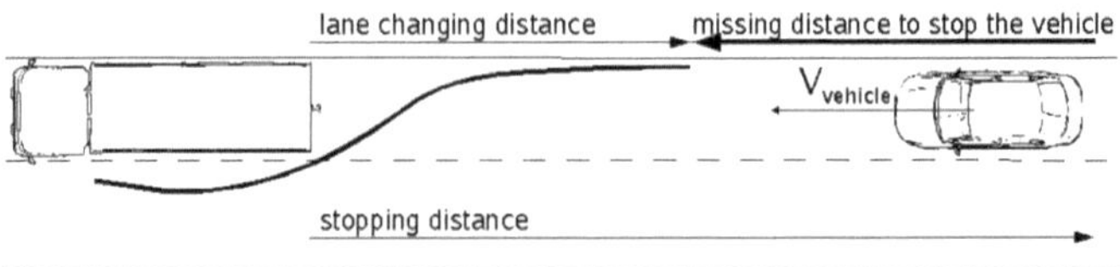

(b) Stopping distance and lane changing for standing object

Fig. 1.1. Risk of initiating the emergency brake too late for standing objects

has time to recover by himself/herself. Unfortunately in another case such as for a standing object, the emergency brake should be started very soon to stop the ego vehicle too.

However, the Vienna convention [35] did not allow the intervention process here. Indeed it is not possible to intervene until the very last moment which means not only with an emergency brake, but also with a lane changing. And unfortunately the minimal distance for lane changing (about 20 m at high speed) will be shorter than the stopping distance. Therefore, it is possible to brake only when it is not possible to change the lane, but unfortunately at this moment it will be no more possible to stop the vehicle. Hence the emergency brake can only be a collision mitigation as described in figure 1.1(b).

Hence the emergency braking alone is not enough to manage standing objects as the lane changing maneuver has to be taken into consideration once the stopping distance has been reached. The full concept with the integration of the lateral control is presented in the following sections.

1.2 Concept of the adaptive cooperation

The cooperation between driver and assistant system is mainly based on the comprehension of the driver's maneuver. Otherwise the assistant system will not cooperate with the driver but it will purely override his/her command, which is not allowed by the Vienna convention. However, in the case of minimal safety level, its action can go down to limiting the driver's command into the safety envelope of the current maneuver.

As already explained before, the cooperation will be done at different levels as depicted in figure 1.2. The real intervention of the virtual driver on the

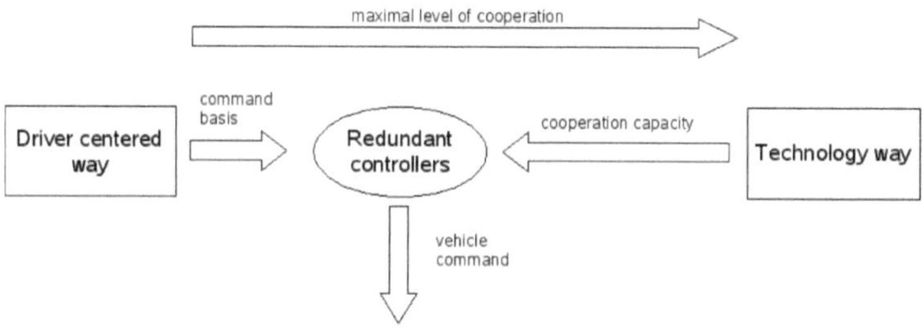

Fig. 1.2. Variable level of cooperation

vehicle command will only be performed on the operative level. Indeed, the capacity of cooperation will be dynamically set depending on two parameters. First the level of delegation required from the driver has to be understood. And secondly the capacity of assistance from the virtual driver has to be defined. Thus, the requirements for cooperation can be checked as represented in the figure below: as long as the assistant system capacity is higher than the requirements, the cooperation can be realized safely. Otherwise, the driver has to know the limit of the assistance, which can be provided to the driver. In normal case, the cooperation from the assistant system point of view can start from pure advices up to a full operation of the maneuver. Furthermore the cooperation algorithm can also decrease the intensity of the cooperation to avoid driver's drowsiness or complacency risks. Last but not least, the driver alertness or capacity has to be known when the system has to degradate the cooperation level. Indeed the decision control needs to verify whether the driver can get back the vehicle control or not.

Therefore the level of alertness from the driver is a fundamental factor. Otherwise, the system can only act with Boolean transition like on current technology but with the extension of a two-dimensional controller. This alertness is the relation between his/her drowsiness and the response to the feedback. Driver alertness (or awareness) is presumed to be necessary but not sufficient for an appropriate focus on external events, i.e., attention or vigilance. Thus, drivers may be alert (i.e., awake) but still inattentive. In the context of driving, "inattentive" means that a driver has failed to perceive a visible crash threat due to "mind wandering," distraction (internal or external to the vehicle) or improper lookout, i.e., "looked but did not see" [128]. The driver information processing error of inattention is widely regarded to be the most frequent principal causal factor in traffic crashes, greatly surpassing loss of alertness [128]. The present distinction between alertness and attention is consistent with past research in this area.

If a conflict occurs on the tactic level (dangerous maneuver or bad transition), the assistant system will intervene with several possibilities. For lower confidence, only advices can be given. However, otherwise mandatory interventions can be performed. However, a conflict is always resulting from a lack

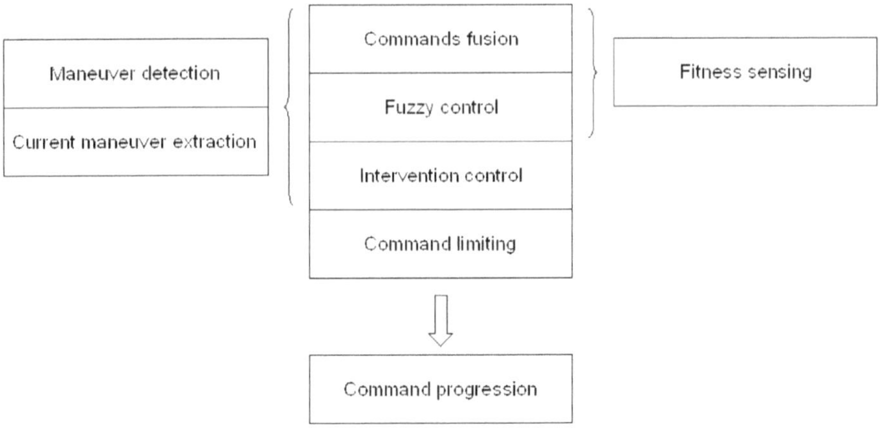

Fig. 1.3. Use of different modules for the degradation of the functionalities

of information either by the driver and/or by the assistant system. Thus, feedback has to be provided to both actors to let them check their cognitions. Such approach is also similar to the concept of Delphi method [78]. Each actor is answers once how to optimize the vehicle command also with a self-estimation. After that, all results are fused together and a feedback is sent to all actors to allow them to correct their answers until the fusion has stabilized. Contrary to the Delphi method, no wrong assessments will be sent to the driver and to the virtual driver to ensure their criticism. Otherwise, it would bore the driver with false-positive and it would use computation time by the assistant system for nothing.

Driver monitoring is useful not only for fatigue detection but also for road scene validation as well. Apostoloff et al. [3] were able to show a clear correlation between the eye gaze direction and the curvature of the road, particularly an apparent monitoring of the oncoming traffic. By monitoring where the driver is looking, many false alarms can be avoided. If the driver is looking at a potential problem/uncertain area in the road scene, a warning is irrelevant. This is based on the higher goal of the cooperation to assist the driver by informing him/her about events, which have not been detected by the driver.

1.3 Functionalities degradation by use of recovery blocks

The module of cooperation is the single link between driver and actuators that can modify drastically the driver's command. Thus, this module has to integrate safe mechanism for the algorithm control to become fail-tolerant. Here, several alternatives of the different algorithms will be integrated within the ECU. So at each time, an adequate alternative can be used when the higher functions are failing or mandatory data are missing, and they cannot satisfy the service. This use of several alternatives gives the possibility to integrate a backward error recovery. The last alternative here will be the reset of the assistant system and the passive transfer of the driver's command to the executive level. The use of recovery blocks for flight control has been introduced by Hecht [58] with a watchdog timer. This methodology is still up to date and will be used here as a last instance for the detection of failures as soon as the worst case execution time of a function has been reached.

After the booting procedure, the alternatives with the highest level for each module will take the control, as depicted in figure 1.3. Before letting the functions to perform thier computations, recovery blocks are used to store the current data. If an exception is handled, it will be possible to re-load this data and to use a degradated function. If this second function cannot realize the same services, the functionalities using these missing results will also have to downsize their level of intelligence. Once a degraded level of the module has been set, it is not allowed to reach a higher level while the functions are not passive, that means when the driver gets the commands fully.

To avoid the runtime overhead due to the primary use of acceptance tests, only the existence of the needed data will be checked before performing the computation. This methodology has been introduced by Randell in [105] to save computation time for embedded systems. Unfortunately no local data can be stored in the several alternates. Actually risks of inconsistency exist as they are not invoked at each cycle. That is why researchers such as Kim [68] get parallel execution of the recovery blocks on different ECUs. Thus, their functions will have to be integrated on slave ECUs if they cannot be integrated without memory (figure 1.4).

Each function inherits the exception handling from the encapsulation class. Internally, the function first checks the validity of its inputs to avoid failure propagation. If an exception is raised, the sending function has to treat the data corruption. In normal ways this function will have the possibility to correct the data after re-computing the outputs. However, here only a

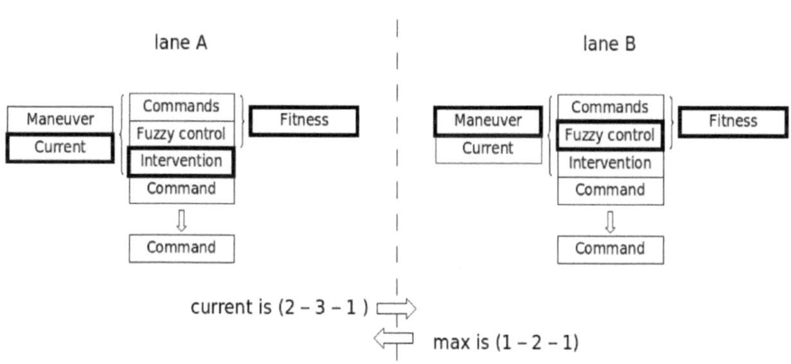

Fig. 1.4. Synchronization of the redundant ECUs

degradation of the function will be performed because of the lack of computation time on the embedded platform. The second exception handling deals with local problem and will try to set back the function on its normal mode by fixing the errors. If it cannot be done internally, the last exception handling will be raised to notify to the service management that the function is unable to perform correctly the request.

There are always some differences in the redundant driver's command because of the sensor noises. Hence, a loss of synchronization may appear if functionalities level is not the same anymore on the redundant ECUs. For example, the choice of the maneuver may differ depending on the way to detect it. Or the algorithm of intervention will realize either a local intervention (adaptive cooperation or fuzzy control) or a full intervention in binary modus. With such differences, the actions performed on the redundant ECUs will diverge. However, an important difference of the output may unfortunately be interpreted as failure from the executive level. Thus, the ECUs have to synchronize their internal states by sharing their functionality levels, the choice of the maneuver and their intervention state. In this case, both ECUs will look for having the same functionality levels by decreasing the performances on both sides. With the same situation handling and the intervention state from the master, the passive ECU will stay automatically synchronized with the other ECU, even if the driver's command already requires an intervention.

After a functionality decrease, both master and slave will try to upgrade their joined functionality level. If no exception raises, the functionality level will be pushed forward again as the punctual exception is over. Otherwise the ECUs will limit the functionality during a few instances, because the source of the problem cannot be determined for the continuous exception.

2

Understanding the driver maneuver

Abstract. The classification of the driver's maneuver will be a fundamental element if the assistant system wants to give adequate advices to the driver. Thus this detection cannot afford neither false recognition nor important latency time. However, the detection of the driver's maneuver is not difficult until a transition occurs. Such a transition is almost done with maneuvers that are quite overlapped with the current one. The single seldom case, which is not like this, happens for a double lane changing. That is why it is possible to reduce the search of the transition to the maneuvers close to the current one. Furthermore this detection has to be fast (lower than 100 ms). Otherwise the driver will feel its latency time and he/she will dislike the not corresponding feedback.

In a different research area, some investigations have already been done. For a few years, detection algorithms have been developed to classify automatically measures on virtual reality platform to create databases. Different methods have been tried but they all work on the sensors level and not on the command level as required here. Pentland and Liu [101] have used a probabilistic approach using hidden Markov chains, whereas Torkkola [127] has used a learning machine.

If this module fails, it will be replaced automatically by a degradation alternative. Its task will only be to determine the maneuver, whose local safety envelope includes the driver's command, and with the closest optimum. The sort will be performed first on the steering angles, and then on the acceleration to get the same kind of maneuver.

2.1 A priori choices by looking at the history

Like every stochastic phenomenon, the driver's choice of maneuver can be determined only by selecting the maneuver with the best probability. In practice the evolution of a dynamic Markov chain, which is depending on an a priori choice and the current command, will be analyzed over the time.

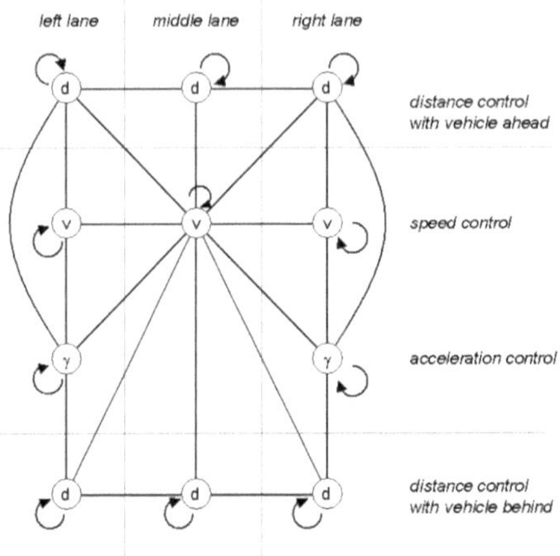

Fig. 2.1. Full chain of transitions of maneuvers

First the history H of the driver is used to define the a priori probabilities of the change of a maneuver i to another j ($P^{\star}_{ij}$) or to itself ($P^{\star}_{ii}$) as in figure 2.1. The history is defined as a probability table dimensions $k \times k$, with k equals to 11. However, some transitions may be disabled over the time because their probabilities are too small. As not every maneuvers are possible at the same time, only a sub-set of the connections will be enabled depending on the situation. The list of the maneuvers is used to disable the transitions that are currently not possible. A transient matrix will computed again with an identity trace by re-defining the probability of variation in order to have a stabilizing Markov chain. Moreover, if some optimums are too close to each other, they will be combined as the driver cannot distinguish them. Here again, their transition probabilities will have to be summed together.

The initial probabilities to stay on the same maneuver are defined by default as 0.5 and the other transitions get uniform value. Step after step this history will be upgraded by using the result of the experiment. The transition chosen will get its probability increased from a step of 0.01 and the other possible probabilities will be lowered uniformly. A minimal probability is set to avoid the risk of having a null probability for the seldom transitions.

The next example presents who from a full chain of transition (fig. 2.1) a sub-set could be extracted (fig. 2.2(b)). The vehicle is currently on a speedway with two lanes between two vehicles. Therefore, on this lane it is possible to control the distance to the vehicle ahead or the distance to the vehicle behind. It is also possible to control the speed as long as it stays on a safe range, which is lowered by the distances to the other vehicles. However, the vehicle can also change the lane to the left, where there is currently no vehicle. Thus for the other lane, only the speed control and the acceleration control are possible.

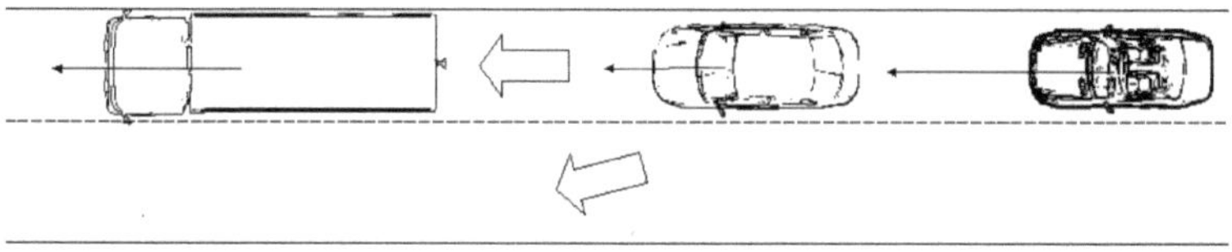

(a) Example of situation

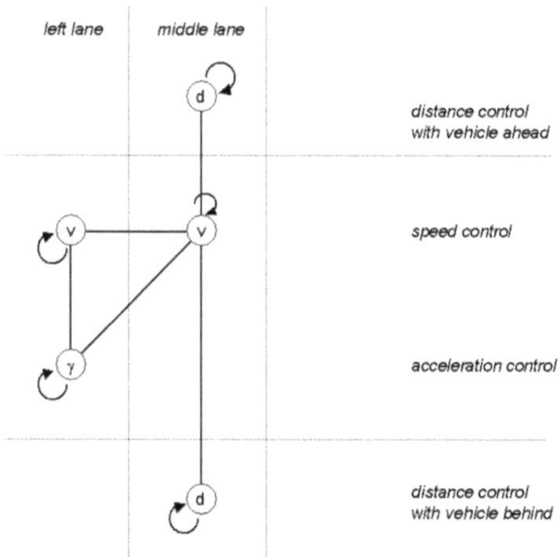

(b) Corresponding local extraction of the graph of transitions

Fig. 2.2. Use of the static graph of the maneuvers for a scenario

2.2 Weighting the choices with the command dynamics

The probabilities of the transition are weighted with additional relevant parameters. The most important parameters to extract the choice of the driver are the distance ($\Delta_{acc.}$) and ($\Delta_{steer.}$) of the driver's command to the optimums of the maneuvers and the variation of these distances. If there is no optimum defined, the middle of the local safety envelope will be used by default. This approximation will lead to a dead time in the recognition if the driver's command stays far away from the optimum but into the safety envelope (the problem mostly comes in terms of acceleration). The first additional parameter is the quality Q of each maneuver. Indeed the higher the quality is, the more probable is the driver's choice because the driver tends to realize an easy maneuver. Finally the state of the indicators is also used to support one direction. The probability of a transition in its direction (C) will grow over the time but after a few second this probability will sink as in figure 2.3. After a few seconds the switch of the blinkers is supposed to be forgotten from the driver and be no more relevant for the computation Fig. 2.3.

The posed problem can be explained as a probabilistic equation to set the probability of a maneuver M_n depending on the former probability M_{n-1} and the properties of the maneuver:

$$P(M_n \otimes M_{n-1} \otimes H_{n-1} \otimes Q \otimes \Delta_{acc.} \otimes \Delta_{steer.} \otimes \dot{\Delta}_{acc.} \otimes \dot{\Delta}_{steer.} \otimes C) \quad (2.1)$$

The equation is split between three independent parts because all the parameters are not correlated. Therefore it is possible to get the probability for each maneuver (i) by using the computation for the Markov chain by using the history H and adapting it depending on the extrapolated positions of the optimums and the qualities as on the next equation:

$$\forall_i, P(M_n^i) = P(M_{n-1} \cdot H) \otimes P(Q^i) \otimes P(\Delta^i) \otimes P(\dot{\Delta}^i) \otimes P(C) \quad (2.2)$$

The probability functions are adapted for each optimum, which means a kernel table per parameters is build. The probabilities of the choice depending on the quality, on the distances, on the variation of distance are all defined as discrete Bell shapes:

- Quality: centerd on 255 (maximal of definition) with a mean value of 100 (dangerous situation). As depicted in figure 2.4, the function permits to disable the maneuvers with low quality.
- Distances: centerd on 0. The mean value is equal to the width and the height of the range for the maneuver. Therefore these curves will get a non-null probability for a motion vector on the boundaries where many maneuvers could overlap.
- Variation of distances: centerd on $5m/s^3$. The mean value is defined by default as $3m/s^3$. In that case the function supports the command synchronized with a maneuver or going into its direction.

This set of parameters has been determined for a small passenger car on a simulator platform in Braunschweig. As for sportive cars, the thresholds corresponding to the variations of command have to be quite high otherwise too much false detections have to be expected. At the opposite heavy good vehicles require lower rate thresholds as their dynamics is really slower.

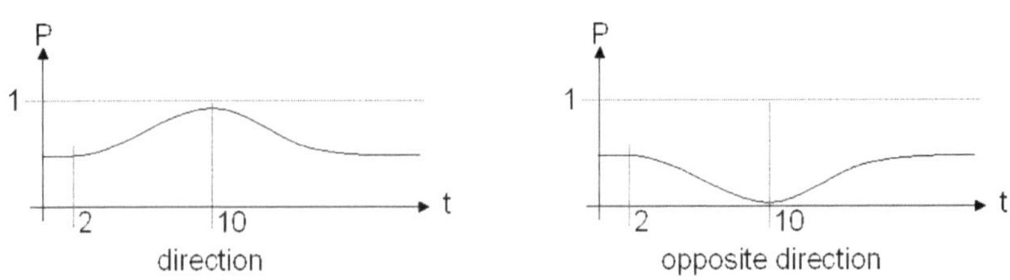

Fig. 2.3. Graph of probability of the directions over the time

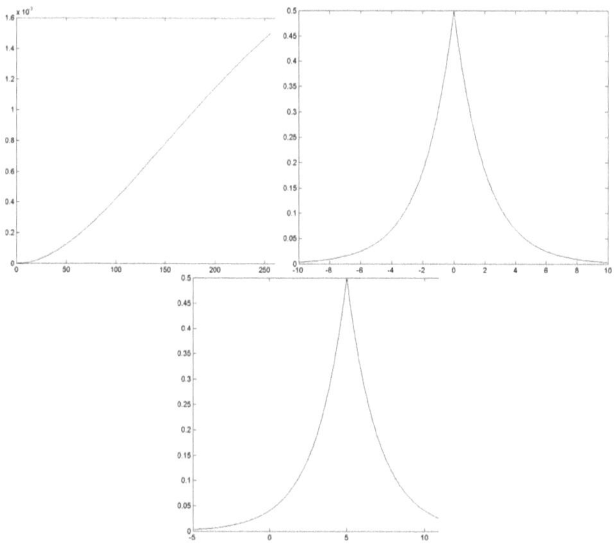

Fig. 2.4. Graph of the probabilities for the quality, for the distances (acceleration in m/s^2) and variations (m/s^3)

2.3 Auto-adaptive detection

2.3.1 Analysis of the probabilistic graph of the maneuver detection

The system of comprehension of the driver's choice gives i probabilities, one per feasible maneuver. However, the driver chose in practice only one single maneuver that he/she will try to realize. Sometimes the driver may hesitate and will set his/her command between both possible scenarios (e.g. before unknown crossing section). Therefore it is really important to find which maneuver the driver tends to realize or to detect when he/she is hesitating. In the second case, the system has to shut down itself automatically and to wait until the driver has chosen his/her maneuver.

When the probabilities for each maneuver are computed, it will be necessary to analyze the result in order to extract which maneuver has been undoubtedly chosen or if the driver hesitates. The repartition of the probabilities is analyzed to extract the chosen optimum if the ratios between its probability and the others are high. Otherwise a low ratio will show an hesitating driver.

2.3.2 Updating the history

At each cycle it is now possible to know if the driver's choice stay the same or change to another maneuver. That implies the possibility to update the history. The history is initialized with a probability of 0.5 to stay at the maneuver and the other transitions get uniform probabilities. Over the time, the history will change until it reaches certain stability. As soon as the history stabilizes, the system will have a theoretical model of the driver. In practice the tests have shown that the history never stabilizes because the driver himself modifies continuously his/her driving manner. Actually the history always tends to a model of the driver that is true only for a few minutes and evolves with the environment and the time. Parallel to this micro-model, a more general macro-model can be also set.

The transition from a maneuver to another one comes rarely on highway. Actually most of the time a driver is performing the same maneuver: *speed control* or *vehicle following*. A driver may sometimes overtake but this maneuver has a short realization time.

The modification of the history has to be done intelligently, otherwise a singular problem will occur. In order to increase the influence of a transition it is possible to set up its probability. By updating every 100ms the history (setting up the transition or the staying and setting down the others) the probabilities of some transitions will surely be once set as zero if the step for the updates is too important. That is why the step for the updates is defined so small that the probabilities will normally not have enough time to go down to zero. Parallel to that, a minimal probability is integrated into the module. As no probability can be null, it is possible to get the probability increasing at each time.

A new approach integrating a concept of entropy may be interesting. By downsizing gradually the entropy of the updating, it will be possible to tend faster to stabilization at the beginning. During the tests, an entropy defined as the inverse of the time has been chosen and validated as first draft method.

3

Determination of the driver drowsiness

Abstract. The term *drowsiness* is used here to refer to the state of reduced alertness, usually accompanied by performance and psychophysiological changes, which may result in loss of alertness or being *asleep at the wheel.* The term *driver fatigue* is also widely used to describe this condition, especially in Police Accident Reports and in accident data files. However, Stem et al. [119], Tepas and Paley [125] and others have correctly pointed out that drowsiness is distinct from physical fatigue and that *drowsiness* rather than *fatigue* should be the principal concern in relation to driving.

3.1 Driver and his/her condition

The problem size of driver drowsiness is much greater for heavy good vehicles as for personal cars because of their driving conditions: long driving time and all day conditions. Moreover the acceptance of measurement devices is much more important for combination-unit trucks than for passenger cars as these systems are more seen as safety tools than as spy [60]. So, various trucks-dedicated devices from small companies are already on the market as reported by Haworth [57].

All statistics show similarity to the typical case of drowsy driver accident. The highest peak is between midnight and dawn, with a second smaller peak after lunch.[1] These accidents happen most of the time in non-urban areas at a speed of around 80 km/h when the driver is alone in the vehicle (75%). More than 70% are single-vehicle roadway accidents. On top of that, these accidents occur most of the time on a straight section (83–17% in a curve) and especially more when they miss the beginning of the curve: most driver did not make any correction to keep the lane. Single drowsy driver crashes

[1] The action of the stomach has a direct influence on the brain activity by modifying the blood traffic.

are strongly related to both age and sex of the driver. In 2005 about 79% of the accidents have involved male drivers in the USA. The same year, 64% of these drivers were under 30 (27% of driver registrations), which makes an involvement rate four times higher than for drivers over 30.

Dingus [33], Vallet [129] and others have analyzed the deterioration of the driving conditions and their results have shown that the drowsiness is always preceded by psychological and performance changes. In addition, Wierwille [136] has shown fluctuation of alertness during long task performance. On top of that, Makeig's results in [81] have shown also irregular fluctuations of these successive times. For drivers, period of low performance are broken with brief lapses of high performance. Micro-sleeps occur punctually without being detected neither by the driver nor by passengers. However, Dingus [32] has shown the possibility to detect them psychophysiologically. However, often drivers still drive even when they are aware of their drowsiness [62].

The first step for the introduction of drowsiness measurement can focus primarily on roadways, when the speed is above 80 km/h. This scenario corresponds to most situations and it is also the easiest case for the measures. Actually, the traffic density in urban situation put in perspective the fall of the driver alertness and it will in parallel introduce noise in the measurement, which will make the estimation more difficult.

3.2 Direct non-obtrusive measurement of the drowsiness

The detection of the driver drowsiness can be performed with reasonable accuracy over direct and indirect monitoring approaches. Direct measures are almost based on psychological performances. The indirect measures are looking more at steering and lateral position fluctuations. Subsidiary auditory tasks can also enhance the accuracy of the detections but they also may annoy or disturb the driver.

3.2.1 Methodology

Two main methodologies are under investigation for the direct measurement. On one side, the whole body is monitored with, for example, measures of hand grip pressure on steering wheel,[2] respiration and heart rates, skin oxygenation, head and body postures. On the other side, brain and eyes can be directly monitored by use of electroencephalograms (EEGs) with various waveform amplitudes (alpha, beta and theta), electrooculograms (EOGs) and measures of eyelid activity with opto-electronic emitters and sensors [119]. The eyes monitoring can provide several measures, like AVEOBS as observer for high-frequency eye-closure measures, EYEMEAS and PERCLOS[3] as measures of

[2] Difficult because this measure can also describe driver frustration.

[3] Percent of total time of eyelids closed 80% or more.

slow eye-closure and NEWDEF, for example, as mixture of both slow- and high-frequency eye-closure measure. The challenge for these technologies is the usability for the driver. For that, these devices have to be comfortable and the less obtrusive as possible without interference with the driving activity.

To lower the problem of acceptance, fully non-obtrusive eyelid monitoring devices are under investigation. With a camera mounted in the dashboard, an image processing unit continuously analyses the driver's face and especially more the eyelid activity (an example of snapshot is presented in Figure 3.1). The inherent problems are the head movements, the partial obstruction from glasses and the varying luminosity. However, additional functionality can be added on top of this functionality. The eyes direction can be also monitored to verify if the driver has seen relevant road-users or shields [119].

3.2.2 Problem of reliability

The driver drowsiness sensing is already used to activate different warning systems. Some guidelines such as the COMSIS publication ([27], [28]) or the deliverables from the European AWAKE project [83] describe roughly the level of activation, the different warning channels and their use. This stimulus must be capable of overcoming sleep inertia [125] but should not cause a startle-response disruption of driver performance.

Thus, the amount of false-positive is critical for the use of this information. Up to now self-calibrated devices can reach up to 0.98 accuracy rate as reported by Knipling and Wierwille [70]. Lot of research such as by Itoi [62] has noticed limitations due to huge difference by peoples. No golden law can be defined: for example, no general threshold can be fixed for heart and blood frequencies, and threshold for eyelid detects drowsiness only in a later stage.

A first possible improvement can base on a two independent stages detection process. One of the algorithms described before can be coupled with a

Fig. 3.1. Direct measurement of the drowsiness with eyes monitoring

secondary task request. The driver will have to respond to demand by pushing buttons, for example, or by answering questions. Analysis of these answers will have to enforce the primary detection to start warnings systems. Another possibility is to shift from psychophysiological performance, which is highly driver dependent, to a driving performance, which can be driver-type independent. Last but not least, additional information such as time of the day or driving time [1] can be used to verify the tendencies. However, this information is not reliable enough for standalone criterion.

3.3 Combination of multiple indirect measures

Up to now, performance indicators have been rarely used, because they have not provided a direct measure of the drowsiness state and could not be used on their own to estimate the drowsiness with a high confidence. Therefore, the new method presented here uses several algorithms in parallel to improve this confidence by considering a multitude of these indicators including their temporal development. Generally, there are several needs an alertness sensing system has to satisfy, as described, for example, in [113] and [135]. Here, the use a drowsiness qualification split into three levels fit, intermediate and drowsy was requested as minimal possibilities with a confidence value.

The time, when a driver becomes drowsy, is very individual. The main focus should be on how the driver handles the vehicle and not when he/she is tired. From the safety point of view this is even an improvement, because not all drowsy drivers are incapable of driving. In this case, a conventional system would initiate countermeasures not necessarily needed. However, monitoring the ability of the driver to handle the vehicle has even more advantages. Not only drowsy drivers, but also persons who have consumed alcohol, drugs or other intoxicating substances can be detected, clearly reducing the driver's performance. In order to establish such an algorithm and for system calibration, test drives from all different categories of drivers are needed to collect data.

3.3.1 Simulation of test drives

In order to develop the drowsiness algorithms as close as possible to the real environment, simulation drives have been scheduled. The simulation runs have been conducted with 10 male participants between the ages of 20–35. This age class has been chosen as it represents the highest risk class. During these sessions it was possible to observe the drivers and record their driving performances. The observation of the drivers is needed as an independent reference for the driver state, in order to prevent faulty system calibration due to the recorded measures.

Each participant passed a short training session of 30 min before the measured test drives. During the training session, the test subject had the opportunity to get familiar with the equipment and the simulation environment.

Figure 3.2(a) and 3.2(b) presents driving performances distributions before and after the training sessions. The hypotheses for a good driving performance are low fluctuations around the mean value distribution as well as low standard deviation values (SD). The translation of the distributions before and after the training to these optimal ranges indicates an improvement of driving performances in simulator.

Every test drive session lasted for 30 min and confronted the test subjects with a daytime a nightly looped highway scenario, which are according to [49] the most dangerous situations for fatigued drivers.[4] Each driver performed three test runs: one in the morning between 8 and 9, one in the afternoon between 14 and 15 and one in the night between 0 and 4. This way a day-time spectrum of driving performances for all drivers has been recorded. The drivers were asked to act as relaxed as possible and to show a normal driving behavior with the goal to stay on the road. During the test drives, no interaction with the driver was allowed in order to prevent eventual discom-

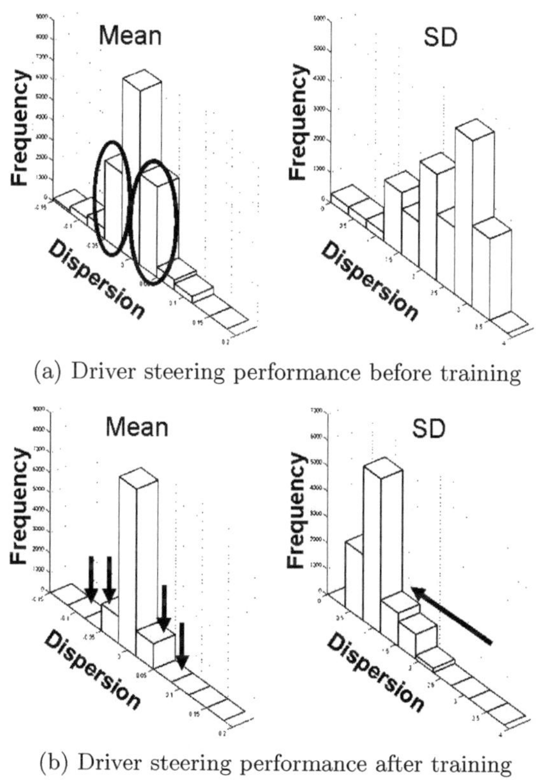

(a) Driver steering performance before training

(b) Driver steering performance after training

Fig. 3.2. Adaptation of the driving skills due to training

[4] In urban situation, the density of actions later shifts the fatigue.

fort or distraction of the driver. At the end, the pilot was asked for his/her personal self-estimation of his/her drowsiness evolution while driving. When the driver's self-estimation of his drowsiness corresponded to at least 80% of the observation, the test drive was considered significant and the measure was kept for subsequent analysis.

3.3.2 From measures to indicators

There are several steps to set up a driving performance measuring algorithm. At the beginning, the desired methods are chosen and their input requirements are identified. Then, with the help of measures collected from test drives, individual modules are defined and tested, upon which drowsiness indicators, relevant measures for the driver drowsiness, emerge. Finally, the results from the modules are gathered and compared. It is clear that a single measure is not sufficient for drowsiness evaluation, especially due to the variety of driving scenarios and drowsiness influences. Even if the measure was perfectly calibrated, a considerable risk of failure would still remain. Therefore, several measures and their reaction to different data sets are observed. Then, using data fusion algorithms an improved and stable drowsiness estimation is computed.

Three different groups of indicators have been identified. The first group concerns all steering-related measures and their derivates. While measuring the quality of the steering movements, drowsy drivers are convicted by their many variations in comparison to fit drivers. The second group characterizes all lateral position related measures and their derivates. In analogy with the first group the hypothesis is made that drowsy drivers are not as able as fit drivers in keeping the lane, having thus significantly increased indicator variations. The last group contains special measures such as the analysis of the time to line crossing, which is predominantly long for fit drivers and rather short for drowsy ones, especially if the curve radius is small.

For the majority of the indicators the main processing is divided into two steps: extraction of the indicator from the raw data, which is followed by a statistical conversion into a histogram representation (figure 3.3(a)). As a pre-requirement for the conversion, a time window, where the indicator is evaluated, needs definition. If this window is too small or too large, no significant indicator measures will be extracted. A practical estimation indicates the window limits to 5 s up to 1 min of driving. Note that the corresponding data amount has to be stored in order to be processed after this time. Thus the choice is equally limited by the storage space at disposition. To meet a compromise, the size was chosen to be the equivalent of 10 s of driving.

As can be seen up to the measure below taken at 10 pm, the indicator values located with a very high frequency in the low zone (segment A): the driver is fit. Furthermore, there is a significant increase in higher indicator values (segment B) until the end of the test drive, where no more low values can be observed but only higher values remain (segment C). This can be interpreted that the driver who was fit at the beginning becomes increasingly drowsy from the 13th measure onward (after 18 min of driving).

On the whole, 16 different indicators have been defined. Note that not all indicators are necessarily needed for the estimation of the driver drowsiness. Reducing the number of indicators has a positive influence on the execution speed of the iterative algorithm. On the other hand the quality of the measure threatens to be significantly degraded, because essential measures are abruptly missing. Following indicators are the most important:

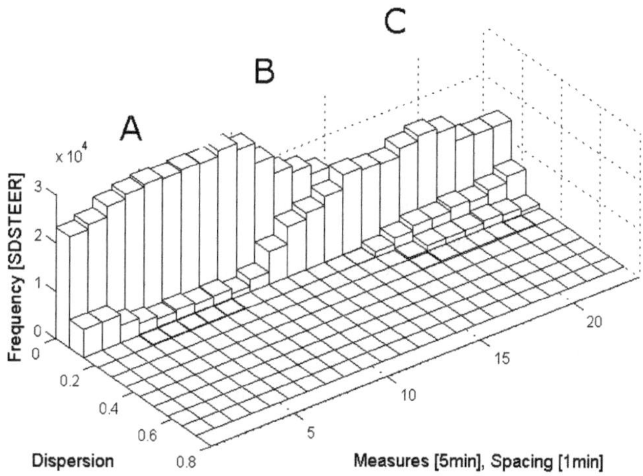

(a) Statistical conversion

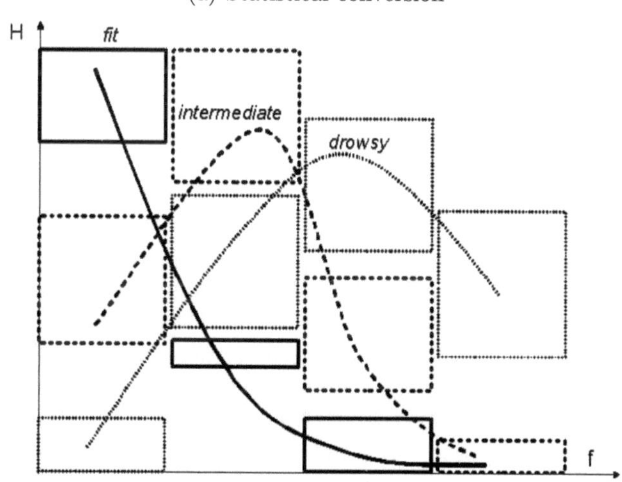

(b) σ confidence of the frequency analysis

Fig. 3.3. Extraction of the models

- *Standard deviation of the steering angle.* This indicator performs an analysis of the quality of the drivers commands. As mentioned before, fatigued drivers have a tendency to enlarge standard deviation values more than fit drivers.
- *Mean of the steering velocity.* This indicator performs a measure of the driver's commands continuity. Here the hypothesis is made that fit drivers have a tendency to slower steering movements because their vigilance and foresight is greater than with fatigued drivers.
- *Standard deviation of the steering velocity.* Low standard deviations point to rather fit drivers, whereas drowsy drivers have a tendency to higher values.
- *Standard deviation of the lateral position.* This indicator performs a quality analysis of how well the vehicle evolves in its track. For fit drivers this value can be expected to be low, because the vehicle is kept in lane with few small driving corrections.
- *Mean of the lateral velocity.* This indicator measures the quality and smoothness of the lateral motion. Fit drivers tend to have low lateral velocities, as they do not need to perform many quick corrections.

3.3.3 Setting up of drowsiness references

Definition of sequences

The establishment of the references needs at least 20 complete sets of comparable test drive data for calibration. The number of bins and the bin size are determined empirically to include even the highest found indicator variations. In a second step, indicator histograms from all test drives have been analyzed and separated according to the categories fit, intermediate and drowsy.

Establishment of drowsiness models

For establishing a reference, the data is grouped by drowsiness type and by bin. Then a mean value is calculated for those bins corresponding to a certain drowsiness type. Simultaneously, a σ confidence interval is set, using the measure which has the maximal distance to the calculated mean. The procedure is repeated for all bins and for all states. As a result, a mean reference model for a specific indicator including a confidence interval is constructed.

Present models never cover the whole possible data space but concentrate on the mean occurrence regions of the corresponding drowsiness type. Measures falling into areas which are not covered within a σ confidence interval see their probability of corresponding to an adjacent state reduced quickly. Transition states will also have a considerably lower probability attributed, as the indicator histograms do not match the reference models entirely.

Examining a constructed stereotype for standard deviation references in general, there is a sort of translation accompanied with an occurrence drop from the fit state reference, having a tight dispersion and a very high occurrence of low standard deviation values, to a drowsy state characterized by a wide dispersion and significantly reduced occurrences. Characteristic for general mean evolutions is the fact that all reference profiles are centered and are similar to a Gaussian distribution having a small σ confidence interval for a fit driver state and a larger σ confidence interval with lower middle frequencies in the case of a drowsy state.

3.3.4 Combination of the drowsiness indicators

The first step in the estimation of the driver drowsiness is the comparison of a particular indicator sample distribution to its reference models. Thus using the frequency of the current distribution compared bin by bin to the defined reference mean and standard deviation, the actual occurrence probability for this state corresponds to the probability that the measure would get in a Gaussian distribution with the defined mean and standard deviation. This procedure is repeated for all indicators and references, which results in a table of probabilities that a distribution in question corresponds to a certain reference state.

Single distribution considerations

There are n probabilities corresponding to the amount of different bins, which estimate a certain drowsiness state. The task of the preliminary indicator fusion is to consider these probabilities in order to extract one representative probability for the corresponding state. Here, the Bayesian formalism, equation (3.1), is used as there are several occurrences for the same event. The secondary indicator fusion repeats the procedure for all different indicators, which generates similar table. Thus the drowsiness states are described by the computed probabilities of the pool of indicators.

Then again, by using the same Bayesian formalism, an inter-indicator probability for a drowsiness state is calculated. Note that the sum of the resulting drowsiness state probabilities is not equal to 100%. This is typical for a system where measures have been made independently, but are linked to each other in reality.

$$P(A \mid B) = \frac{P(A) \cdot P(B \mid A)}{P(A) \cdot P(B \mid A) + P(\neg A) \cdot P(B \mid \neg A)} \tag{3.1}$$

Global state fusion

The final drowsiness state is estimated using Markov chains. For a given starting configuration of drowsiness state probabilities, the final state is determined by the exploration of the convergence of the Markov matrix. Beforehand, the transition probabilities required for the formalism are determined by use of a connection graph between the different states. The law between the states is simply taken as the inverse of the class distance: 1/2, 1/3 and 1/6.

$$\begin{bmatrix} P_{fit} \\ P_{med} \\ P_{drs} \end{bmatrix} = \lim_{x\to\infty} \begin{bmatrix} p_{fit,fit} & p_{fit,interm.} & p_{fit,drowsy} \\ p_{interm.,fit} & p_{interm.,interm.} & p_{interm.,drowsy} \\ p_{drowsy,fit} & p_{drowsy,interm.} & p_{drowsy,drowsy} \end{bmatrix}^x \begin{bmatrix} 1 & 0 & 0 \\ 0 & 1 & 0 \\ 0 & 0 & 1 \end{bmatrix} \quad (3.2)$$

3.3.5 Following the drowsiness evolution

From the total effective of 60 recorded simulation drives a correspondence of the self-estimation of the drivers to the observed drowsiness was established in 90% of all drives. A verification of these subjective impressions was then provided with the help of the drowsiness evaluating algorithms.

Driver self-estimation versus computed drowsiness

The plot of the self-estimation of the driver can be superposed to the drowsiness graph to verify the reliability of the measures. The self-estimation consists of three values, one for the estimated drowsiness state in the beginning, one during and one at the end of the test drive. It is possible to see that the estimation in the beginning and at the end differ with an error lower than 10% to the calculated drowsiness. For the intermediate estimation the error is approximately 15% during 1000 s before dropping to lower values. Note that especially in the transition states of the drowsiness around the intermediate estimation, the algorithm faces shifted histograms which cannot be clearly attributed to a certain drowsiness state. These result in a delayed follow up of the computed drowsiness to the actual estimation causing a slightly increased error.

Figure 3.4 shows the situation for the example drive, where all indicator measures are activated. Immediately it can be seen that in the beginning of to the 20th minute (1200 s), the driver was estimated very fit with a very high probability. From then on, the drowsiness increased quickly accompanied with a slight decrease in confidence.

Extending this consideration to all recorded measures, an average error percentage for the correspondence of the self-estimation to the drowsiness algorithms leads to the conclusion that the drivers tend to misjudge their drowsiness state especially in the starting phase of the driving simulation, which is why an auto-calibration of the algorithms based on the driver self-estimation is hardly possible. With time increasing, the error decreases because the drivers self-estimation becomes less differentiated. In the majority of all cases, the indicated self-drowsiness is simply a fatigued state or a very fatigued state, possible to match easily with the computed drowsiness.

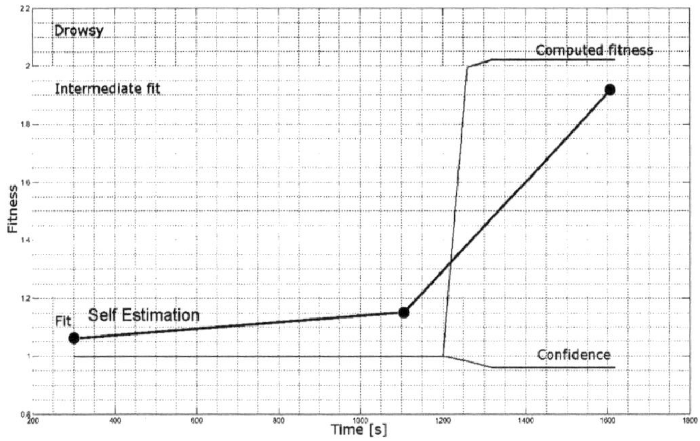

Fig. 3.4. Correlation of the driver self-estimation with the estimated drowsiness

Indicator performances

The statistical result produced by the indicators on the reference sequences are evaluated using a Pareto similar analysis (see also [44]). It allows identifying quickly the most important parameters for a certain effect. Thus for the state in question, the probabilities of the indicators identifying this state are extracted and sorted in a descending order. In the next step a cumulated probability graph is plotted. Figure 3.5 presents the cumulated conditional probabilities for a true-positive of fit state. The graph is divided into three segments with different probability ranges.

Up to two-third of all indicators, the identification probability is superior to 0.5. This signifies that a large majority of the indicators, namely those contained in this segment, are able to identify the driver drowsiness state

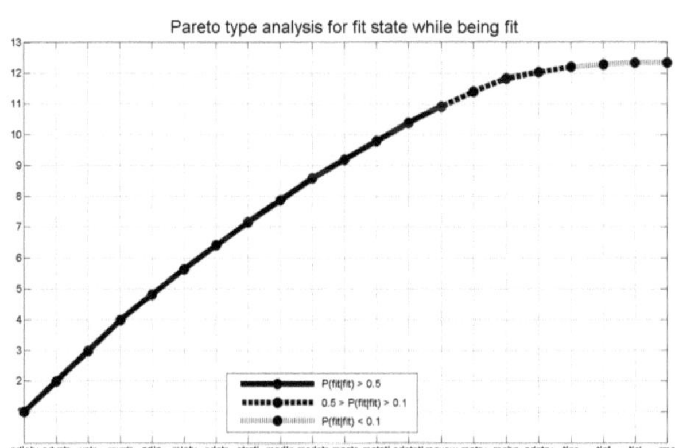

Fig. 3.5. Indicator performances detecting a fit state, the driver being fit

correctly with high probability ($p_{state} \geq 0.5$). In the dashed segment there are indicators whose identification performances for this drowsiness state are inferior to 0.5. In the dotted segment the probability decreases even below 0.1. Thus it is possible to conclude that the indicators located in the latter two segments serve not as efficient fit state identifiers. There are two explanations for this. Certainly the statistical reference model of the drowsiness state for this indicator is not optimal and needs an adjustment which would cause a probability increase for the indicators concerned, or eventually the indicator acquisition also needs modifications such as noise reduction or compensation.

4

Cooperation at the command level

Abstract. In this chapter only the three algorithms with the highest level of intelligence will be described: binary intervention, fuzzy substitution and adaptive cooperation. Actually the first method which is based on command limitation does not present any technical difficulties. Moreover the experimental results about driver behavior with this technology are corresponding to the analysis performed formerly with flight controls: a high risk of complacency cannot be avoided.

4.1 Binary intervention

4.1.1 Concept of intervention

In the case of a binary intervention, the vehicle command is always corresponding to the driver's command until the intervention control has to intervene. This intervention will happen at the very last moment before the driver's command goes through the boundaries of the safety envelope. At this moment, the intervention control will use the corresponding maneuver optimum to optimize the maneuver that the driver has tried to perform. Such optimization will be done as long as the driver's command cannot be sent again to the vehicle command without improvement.

In parallel, the feedback to the driver has to be provided correctly parallel to the optimization process. An adequate feedback will help the driver to perform a similar correction to the intervention control even if he/she does not have any influence on the optimization process. It will have several positive influences such as the short time of intervention, minimization of the complacency risk etc. Nevertheless if this feedback has not been performed correctly, a dead lock would happen.

Indeed if the driver does not understand what is happening and thus he/she does not correct his/her command, the intervention control will stay in optimization stage for ever. Once the maneuver is performed correctly it will be transferred automatically either to a distance control or to a speed control for the current lane. So, the dead lock can be avoided by using a timer which will transform the maneuver to a control of the current speed which will decrease over the time until the vehicle stop. Furthermore when the driver's command does not require an optimization anymore, a second fast transition will be performed from the current vehicle command to the driver's command.

When the decision control has to improve the driver's command, it has to realize a transition from the driver's command to the optimum as fast as possible. However, it does not imply automatically a direct transition, which can go through the safety envelope and provoke an accident instead of avoiding it, as in figure 4.1. Therefore a transition path has to be computed with three requirements: the fastest with the fewest modifications of the command and always being far away from the safety envelope.

The preparation of the intervention has to be performed during the first cycle time of the intervention mode within 1 ms. Hence the level of complexity has to stay as low as possible. Among all the existing methodologies, a path planning based on a grid has been chosen for its low complexity. The algorithm generating the meshing shall deal with all kinds of safety envelope. Thus, a method based on finite elements with Delaunay method [26] may be the best solution. However, this algorithm will be drastically modified to avoid having any recursions (risk of stack overflow and saving time overhead) and to work with non-convex forms.

4.1.2 Meshing algorithm

Up to now the safety envelope is still defined as a group of sub-safety envelope which corresponds to each lane. In each sub-part of the safety envelope up to three optimums may exist: speed control, distance controls to the front and to the back. Thus the direct meshing algorithm will use the sub-part of the safety envelope as exterior edges and the optimums as vertices as shown in

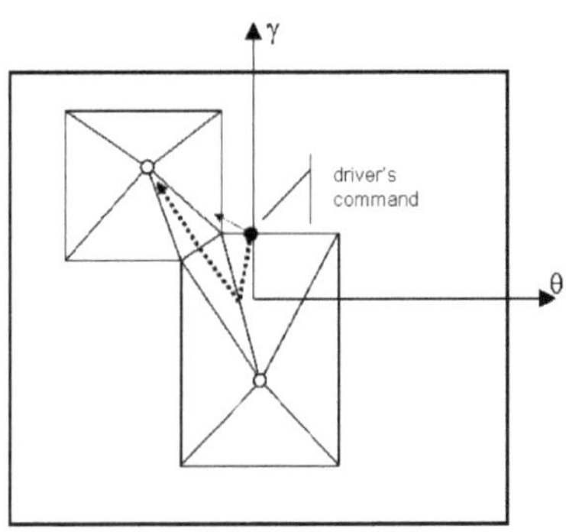

Fig. 4.1. Use of a Delaunay's algorithm to find a path

the figure below. With this limitation of points, it will be possible to update automatically the meshing as long as the overlaps do not change (Fig. 4.2).

If there is no intersection with other sub-part of the safety envelope, the meshing algorithm will be performed in three steps:

- Creation of the first triangles as in figure 4.3(b). In case of intersection, two optimums may have the same acceleration. In such cases the closest optimum to the middle of the steering range will be used.
- Creation of the internal triangle if three optimums are defined. Otherwise the segment between the two optimums will be used as internal edge of the last triangles.
- Creation of the last external triangles

The constraint used in the Delaunay's algorithm to create a triangle is based on the non-inclusion of other points into the circle generated by three points as triangle. As long as this constraint will not be replaced by an easier one, it will not be possible to set directly the triangles. Thus, a sufficient but not optimal constraint will be used instead of the original one: the average maximization of the triangles' surfaces.

Two cases are still possible after the internal triangle has been set. Let us look at this triangle. On the opposite side of the vertex in the middle, there are only two possibilities of triangles as depicted on the left side of figure 4.3(d). This case corresponds also to both side of the segment created when only two optimums are defined. Here an exhaustive computation of the four triangles can be performed to choose correctly the right meshing.

On the other side, the amount of possibilities is much higher. Moreover some edges may not be possible if they cross the internal triangle. However, the set of the triangles can be performed anyway. To minimize the differences between the triangles' surfaces, those triangles have to be optimized locally by means of being upper and lower than the optimum in the middle. A mathematical proof can be given by an exhaustive computation of the surfaces.

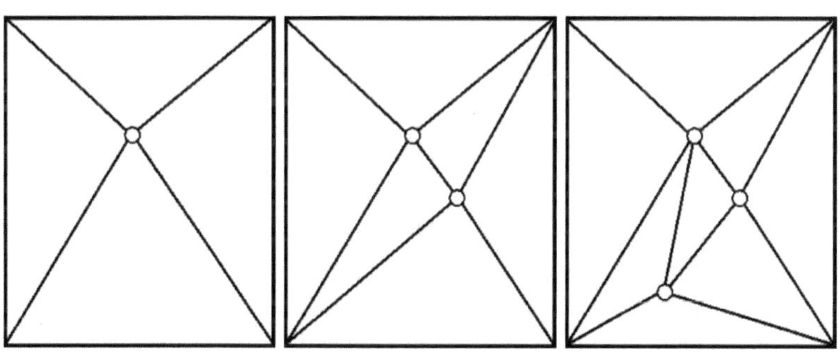

(a) 1 optimum (b) 2 optimums (c) 3 optimums

Fig. 4.2. Results of meshing with different amount of optimums

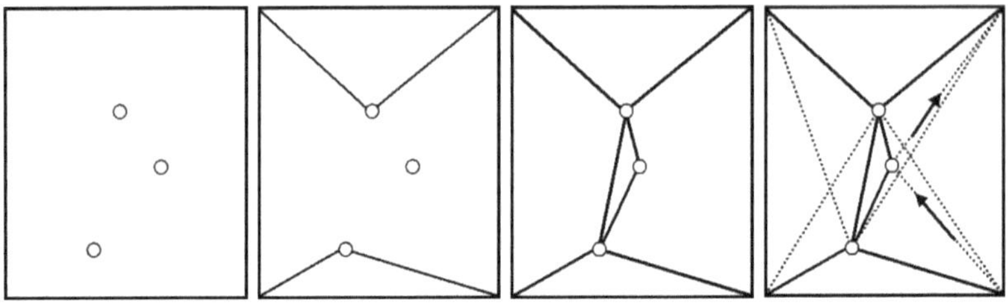

(a) Set of the optimums (b) First triangles (c) Internal triangle (d) List of the last possible triangles

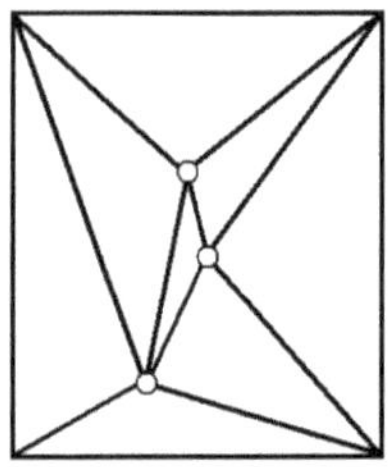

(e) Set of the last triangles

Fig. 4.3. Generic five-step meshing algorithm

Indeed this local maximization can be done automatically by choosing the triangles as in figure 4.4(a).

The limits setting the last triangle can be created on the fly by choosing the zigzag line joining iteratively the vertices from the sub-part of the safety envelope (as in figure 4.4(a)) and the optimum in the middle. Normally there are always two possibilities but both cases will not be computed to save time.

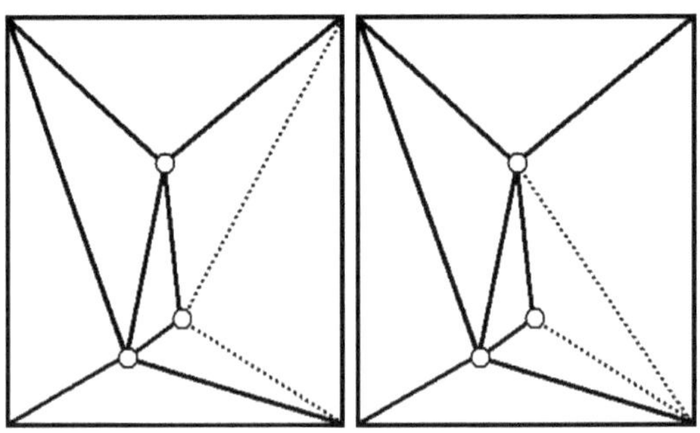

(a) First possibility of meshing (b) Second possibility of meshing

Fig. 4.4. Results of meshing with different amount of optimums

Even if the automated choice starting from the lowest vertex of the sub-part of the safety envelope is not the optimal one, it will be safe enough to manage the intervention anyway.

This algorithm can also be extended to the cases where sub-parts of the safety envelope overlap. By splitting the gap as in the next figure, each sub-part of the safety envelope can be meshed independently with the unmodified algorithm. The single modification deals with the creation of the first triangles which may integrate now the apexes generated by the intersection (Fig. 4.5).

4.1.3 Computation of the path transition

First of all, the way to describe the list of connections between the different apexes A has to be fixed correctly otherwise reading/writing time overhead will occur. When the amount of vertices V is dense, that means proportional to A^2, a matrix $V_{i,j}$ can be created to define the connections and the distances together between an apex i and an apex j. In such cases the initialization of the matrix does not create an important time overhead. However, here the upper bound of the amount of vertices stays in a low range where it can be accepted as quasi-linear [1]. Therefore, a better way should be to define only the few connections by using dynamical graphs. Such a programming way is not acceptable for embedded system with high criticism which needs to have a deterministic computation time. Thus, a static table of connections has been set implemented with an upper bound having an adequate buffer.

The weighting of the vertices is depending on the distance between two apexes. Unfortunately the units are not homogeneous: acceleration is in m/s^2 and steering angle is in rad. No weighting has been done and the distances have been taken as given because the transformation of the steering angle

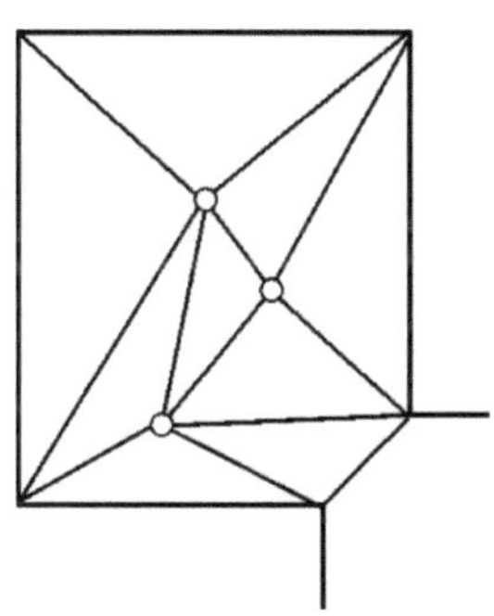

Fig. 4.5. Add of a segment to split a gap

[1] 5 apexes produce 8 vertices, 6 apexes produce 11 vertices and 7 apexes produce 14 vertices

units to lateral acceleration requires too much computation time with a low influence on the results.

The second step is to define the starting point of the graph and the goals. The starting point is clearly defined as it corresponds to the triangle including the driver's command. As an optimum can be used several times as apex for a triangle, more than one goal can be defined like in figure 4.6. Thus, the possible goals will be marked by use of an additional flag in the table. In practice the algorithm will tend to the closest goal.

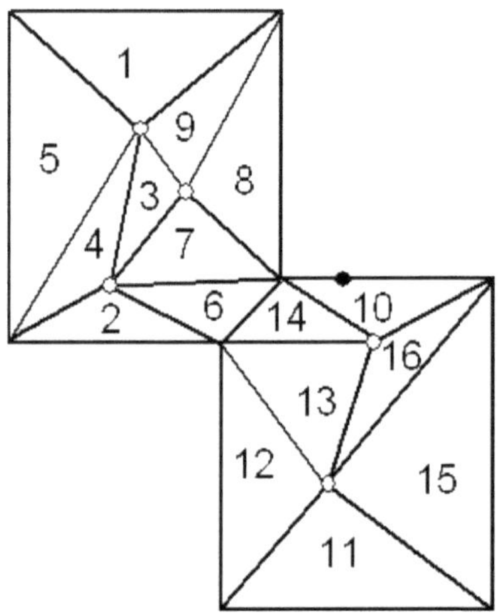

(a) Set of the starting point and the different goals

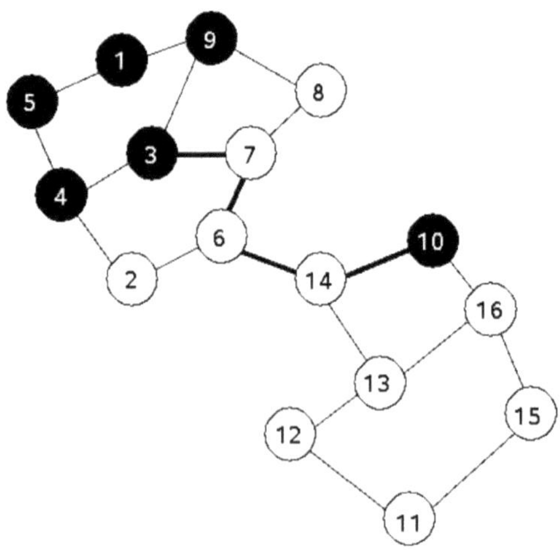

(b) Corresponding graph

Fig. 4.6. Intervention to optimize the overtaking maneuver

Table 4.1. Table of closest triangles for a path transition

10	14	16	6	13	15	2	7	12	11	3	4	8	5	9	1
14	6	10	2	12	11	4	3	11	12	4	2	7	1	1	5
16	10	13	7	14	16	6	6	13	15	7	3	9	4	3	9
	13	15	14	16			8			9	5			8	

The algorithm can now determine the different sequences to take, depending only on the connection graph. This algorithm is presented by taking the current example shown in the former figures. The transition path will have to reach triangles 1, 3, 4, 5 or 9 from triangle 10. The first step is to establish a list of all possible reachable triangles from the first triangle 10, by using the table as shown below. In this example, triangles 14 and 16 are neighbors of the start triangle 10 (direct connection) with their corresponding distances. Therefore triangle 14 is the closest neighbor triangle of 10. It is impossible to find a way via the triangles 16 from the starting triangle which is shorter, because the distance to these points is already longer than the distance between triangles 10 and 3. The table presented before summarizes the links. The procedure will be done recursively until one of the final triangles is reached. Here triangle 1 is the 16th nearest triangle of triangle 10 (last column of the table) in terms of apex connections.

These triangles can be sorted after that depending on their distance to triangle 10. Unfortunately, it will not be possible to sort them directly by summing the length of the different sections to reach the destination because of loops which would be generated. Thus an iterative test of closest distance has to be performed. To determine the shortest way this grid will be read from triangle 3 (closest triangle connected to the goal) back to triangle 10. First the neighbor of triangle 3, which is the nearest from triangle 10, is determined by looking at the vector of motion. In this example, triangle 3 is represented for the first time on the column for triangle 7. Now the next triangle on the shortest way from triangle 3 to triangle 10 corresponds to the triangle in the column where triangle 7 is represented for the first time. This is triangle 6. If this operation is repeated until getting triangle 10, the shortest way will be determined as the path going through triangles 14, 6 and 7. Furthermore in practice the transition path will get through the middle of the common vertex of two consecutive triangles to be as far as possible from any limit.

4.1.4 Transition control

As soon as the vehicle state reaches the safety envelope (depending on the actuators dynamics) the intervention control will act to avoid the accident. In this case, it has to transfer the vehicle command from the driver's command to the optimum of the chosen maneuver (or vice versa) as the power train cannot jump from one command to another completely different command. At this moment the transition path will be used and realized step by step Fig. 4.7.

A controller 1 gives intermediary milestones from the path transition with constant distance which is close enough to be considered as continuous. Furthermore, this controller is supported by another correcting controller to have different points of command between these milestones. A second controller is used later to set the vehicle command adequately in order to realize the needed transitional state.

The command calculated by the first controller is updated at every sample time with a report of distance. Furthermore, each point of the path and of the controller permits to define a range which induces the change of the command given to the power train. Indeed, the strict reaching to the different points of command is useless to conduct the driver to the wished optimum.

The stability of the controller will be ensured by using the transfer functions got from the power train as actuator model. This controller split the corrective action in terms of acceleration, braking and steering; and for each one it uses a parallel transition controller which uses the corresponding transfer function. The parameters of the PID controller are taken as the parameters of the inverse of the given transfer function. On top of that, an integration part is added automatically to prevent the controller diverging.

4.1.5 Critical analysis

The algorithm presented just before gets two drawbacks. Actually, the main hypothesis is that the limits of the safety envelope are known with accuracy higher than the discrete step. Thus, on one side the positions of the objects are supposed to be always well known and on the other side, the vehicle dynamics is perfectly modeled even when stochastic disturbances are due to the environment. As both hypotheses are not true during all the time, the credibility of the safety envelope may depend on the situation. That means, the intervention control may not always have the right to take the initiative (Fig. 4.8).

It is possible to decrease the risk of false actions by downsizing automatically the safety envelope to have a safety range against all uncertainties, or by disabling the intervention process as soon as the context is not 100%

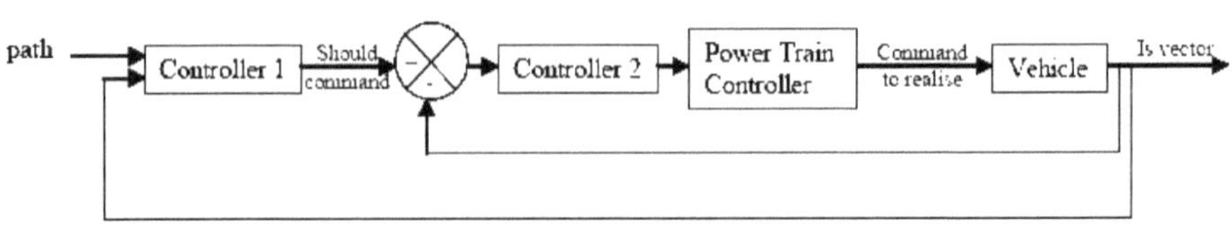

Fig. 4.7. Controllers for the transition

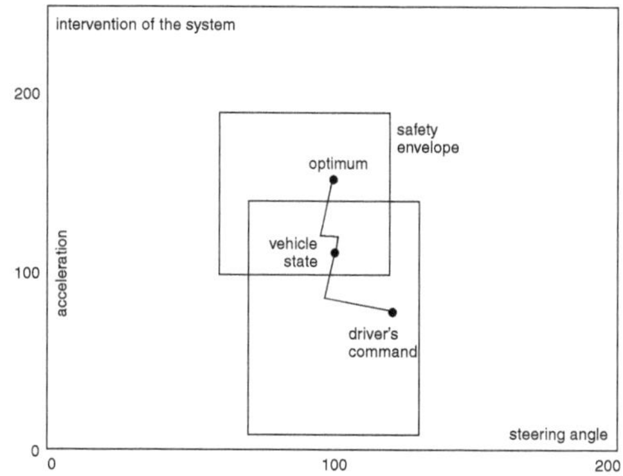

Fig. 4.8. Results of a transition

under control. However, the first possibility will may frustrate the driver—especially the sportive drivers—because it would decrease their degrees of freedom. Moreover even an automatic decrease of the safety envelope will not ensure a safety concept as objects may not have been seen for example, which leads to totally different safety envelope geometry. On the other side, if the intervention is disabled as soon as any uncertainty arises, the system will not be used with all its capacity at all. It will be used only on some dedicated easy cases (e.g. short range dead angle) and not on complex and dangerous situations, exactly when the driver needs support.

4.2 Fuzzy control

One possibility to forecome this problem is to take into account the confidences of both driver and virtual driver. Thus the level of substitution can be weighted by the risk of false action. The fuzzy substitution presented here will determine the weights of both driver and virtual driver to have the vehicle command attracted by the most confident one. With such approach the source with the highest confidence will have the opportunity to overcome the influence of the other actor: a confident driver can override the assistant systems and vice versa.

4.2.1 System confidence value

The confidence value of the virtual driver is directly the product of the confidence value of the sensors and of the agents because the two confidences are independent by use of the data fusion. From the sensors several confidence values can be defined as for each data there is a corresponding probability. Thus the confidence value has been set by default equal to the probability which has the most influence on the situation patterns of the agents, as e.g. the probability of the distance of an object ahead. At the opposite, the confidence value of the agent is corresponding to the matching rate of the pattern with the real model. Here again, only the most influencing element will be taken into account such as the difference between the distance of the vehicle ahead and the distance range in the pattern for the same situation.

Furthermore the fitness for the driver has also to be correlated with the acceptance of the feedback to describe his/her confidence value or better named the activity. This information is required for the operative level to weight the level of intervention from the virtual driver. Furthermore, this information is also valuable during maneuver transition to help giving adequately a feedback relative to the tactic level.

4.2.2 Adaptive weighting

On one hand the decision control gets the driver's command and information about his/her fitness. On the other hand the system gets the confidence of the computation, the safety envelope and sometimes the corresponding optimum.

Let us define the confidence of the driver as $c\%$ and the confidence of the virtual driver as $c^{\star}\%$ to fuse the driver's command and the optimum of the maneuver together. A method based on a Markov chain will be used, that can also be extended to n commands (like two pilots and an assistant system in an airplane). Each element will sort the group depending on his/her quality. For the driver, the a priori quality is c and for the other an equal probability: $(1-c)/(n-1)$. This delivers here the a priori matrix (M), which is always transient. Therefore the real weights of each element can be computed as the stabilization of the Markov chain:

$$\forall i \in \mathbb{N} \times [1, n], [P(E_i)]_n \cdot O_n = \lim_{x \to \infty} M^x \cdot I_n \tag{4.1}$$

If a transition is detected, there will be anyway no disruptive position change of the optimum allowed. Otherwise the fusion of the command will introduce also a jump for the vehicle control and this is not required. Therefore the optimum will be substituted with a temporary command during a transition. This temporary command will realize the transition path between the two optimums by reusing the method explained on the last section.

This transition path is computed in step with the maneuvers' limit and the optimums, so it depends on the virtual driver. If the virtual driver is not supposed to be pertinent, the path must not be supposed to be true too. Therefore, if the probability of the virtual driver tends to 0, the path between the two commands is a straight line. That is why the vehicle command is calculated in step with the distance between the two commands and the distance between the transition path and the straight line as depicted in figure 4.9 (b).

So the principle of the algorithm is to weigh the two commands on the straight line, in step with the probabilities and then to calculate the center

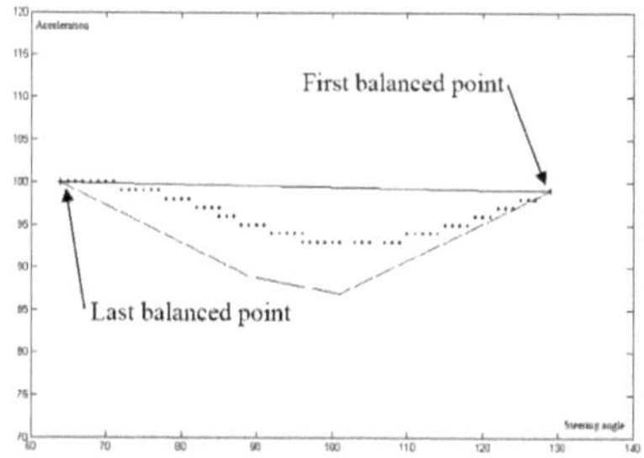

(a) Balanced command during a full transition

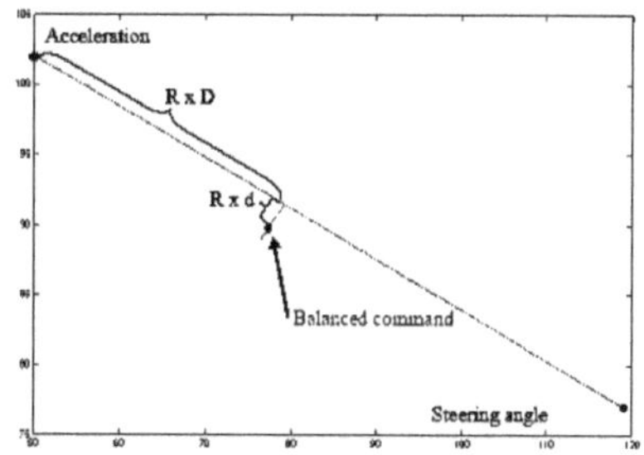

(b) Position of the fusion of the commands

Fig. 4.9. Variation of the intervention by modifying the confidence levels over the time

of gravity on the perpendicular line at the first point, cutting the transition path.

If the driver falls asleep or the confidence of the virtual driver goes down to zero, the vehicle command will go from one element to the other because the gradual modification of the confidence values will modify the influence of the elements as shown in figure 4.9 (a). For example, when the virtual driver takes complete control of the vehicle, the vehicle command is near to the optimum of the chosen maneuver. However, the fall of its confidence value makes the driver's command catching the vehicle command. Parallel to this attraction, the transition path will get less and less capacity of attraction on the vehicle command and indeed the vehicle command will tend to go straight between both the commands.

Finally if both confidences are going down to zero, the vehicle will have to brake with the maximal braking value of the current maneuver in order to avoid accident with high energy.

4.2.3 Critical analysis

The fuzzy control method has still one drawback. The intervention process is realized at each time even when the weight of the virtual driver can be set to zero by necessity. A last improvement will be to integrate the dangerousness level of the situation and the level of intervention required from the driver.

4.3 Adaptive cooperation

The binary intervention presented before is actually mixing two different procedures. First an intervention is performed as soon as the command reaches the boundaries of the safety envelope and after that a second process improves the vehicle command by reaching the maneuver optimum. After that the fuzzy intervention had extended the capacity of the substitution process depending on both levels of confidence. However, the last improvement to evaluate is dealing with the adaptive fusion of the two processes.

4.3.1 Concept of accepted dangerousness

If the system has to make a difference between the process of substitution and the process of intervention, it has to be aware of the dangerousness of the situation. Indeed, a higher difference between the driver's command and the maneuver optimum may be accepted when the situation is less dangerous.

Unfortunately the definition of the dangerousness is quite ambiguous as a situation can be described as dangerous or not depending on the driver's mood and knowledge as already shown by [120]. However, one norm has to be defined independently from the driver. Thus with a same level of dangerousness, a situation will require more or less substitution or intervention depending on the driver's state and the feedback will be adapted. Otherwise, a dangerous level adapted to the driver and the current instant will automatically link the dangerousness of the situation to the driver activity. In this case, the system will not be able to choose correctly the type of support like for the fuzzy substitution.

As neither the time to collision nor the time to line crossing can be obtained from the safety envelope, another way has to be found. In practice, the surface of the safety envelope will be used here, because it describes directly the amount of discrete possibilities of vehicle command, which corresponds to a risk of collision. The calculation of the surface of the safety envelope is easy as the meshing algorithm described before provides already triangles.

However, the calculation of the surface is not sufficient because it does not take into account the confidence of the creation of the safety envelope. An accepted dangerousness will be defined depending on both dangerousness and confidence level: a lower dangerousness with a higher confidence level has to influence like a higher dangerousness with lower confidence. Thus, a look-up table presented in figure 4.10 has been used here.

When the size of the safety envelope goes down to zero, a collision will not be avoidable. Thus, this envelope size can be used to trigger the pre-crash function. However, the crash point cannot be extracted.

4.3.2 Extension by use of the accepted dangerousness

If the dangerousness of the situation can be taken into account now, it will possible to discern two different cooperation cases at the boundaries of the dangerousness range. On one side, a substitution process can be used when the situation is not dangerous. Such a substitution process will deal with the optimum of the maneuvers. On the other side an intervention process can be started to prevent an accident. These two processes are not incompatible because it is possible to require an intervention even with low dangerousness,

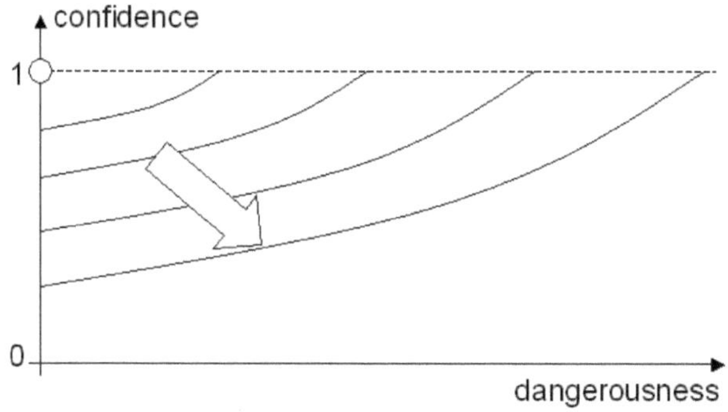

Fig. 4.10. Look-up table to define the accepted dangerousness

and at the opposite the assistant system can be substituted to the driver during dangerous situation. However, each process is mostly focusing on one dedicated case. However, this overlap will make the complacency risk lowered as the use of the substitution process may be triggered to avoid such problem.

Even if the dangerousness of the situation is known, the assistant system will not perceive how to cooperate with the drive at all. Effectively, the level of cooperation required from the driver has to be taken into account. Thus, the level of activity will be monitored continuously. Indeed, an active driver will request less cooperation because his/her own confidence is high enough to be valuable. Thus, the level of the substitution process can be decreased for example, from active substitution down to purely informative feedback. On the other side, a driver with low activity will require a higher cooperation. Here the intervention process, for example, will have to start earlier to the accident prevention. Unfortunately the levels of cooperation are discrete. The lowest level is only visual information, then auditive and then haptic feedbacks will be used and finally a physical access to the vehicle command will be provided. However, a continuous effect can be defined anyway by varying the range of use of each feedback channel. This variation will give the impression to the driver, that they are more than only three steps.

Furthermore a third parameter has to be taken into account. Actually the level of cooperation described just before is only corresponding to the request from the driver side. On top of that, the capacity of cooperation from the virtual driver will be checked. As long as this capacity of cooperation is higher than the requested level, the virtual driver can adapt its strategy to this request. However, when this request is too high for the virtual driver because the situation may be too complex or the perception range might be too small, the cooperation level will be limited to its maximal capacity. In such cases the driver will have to understand that he/she has to take his/her responsibilities and cannot rely only on the assistant system. Furthermore, the risk of complacency will be managed here easily because the algorithm can inverse time to time its reaction to the driver activity in order to have him/her re-integrating the control loop instead of expecting any substitution. However, if the driver cannot come back to the control loop, the algorithm can realize the substitution process anyway.

In figure 4.11, the influences of those three parameters are depicted. On both extremities of the dangerousness range, the two processes are defined. On the other axles the driver activity level is represented from full passive (or shrinking) up to full active. The substitution process can be stronger when the assistant system gets a high confidence level, which means it can act on the vehicle command even when the driver has low activity and requires a substitution. The intervention process will also have the same kind of reaction. With a low confidence level, the assistant system will intervene later on and with a lower level of influence from improvement of the vehicle command down to pure information how the driver may correct his/her command. Thus, in the given figure, the curve of iso-confidence level of the assistant system always decrease with the dangerousness level and with the increase of the driver activity: when the situation becomes to be dangerous, both sources have to be sure enough before acting. The assistant system can perform a mandatory intervention anyway with a high-enough confidence level.

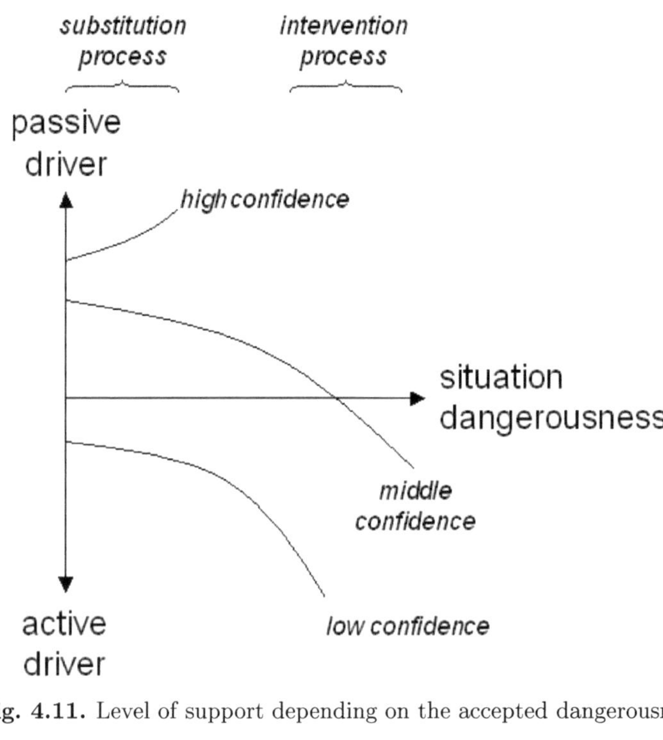

Fig. 4.11. Level of support depending on the accepted dangerousness

4.3.3 Goal-based substitution process

This process is mostly dealing with the optimization of the vehicle command by looking at the optimum of the maneuver. Depending on the confidence level of the virtual driver, the substitution process can be more active or not. Three main levels of activity can be set in advance: informative level, intensive feedback level or even correction level. The two first levels have no access to the vehicle command as they are more related to the tactic level and not to the operative level. Even if the system can bring an important substitution level, it has to determine which level is required from the driver. Therefore an analysis of the driver response to the stimuli (on the steering wheel or the pedals position) is performed with increasing or decreasing steps until it stabilizes. Such converging approach can be used not only to define the acceleration required from the driver, but also to define the speed or distance request.

The concept is based on the torque given to the driver as feedback. Up to now, sportive vehicles have harder feedback to the driver because of the hard damping control. At the opposite sedans have smoother damping control which shows to the driver that his/her command is more adapted to the situation. Such kind of feedback is interesting for the driver as he/she can feel continuously what the vehicle is doing. Here again such feeling will be provided. When the driver gets the vehicle command without intervention, the feedback as torque will be harder and quite close to the real unfiltered torque. However, the intervention of the virtual driver will generate a smoother torque feedback to show how far the intervention is performed.

The fuzzy intervention developed before can be re-used here for the third level of substitution (where the system can also intervene on the vehicle command) with the weights depending also on a substitution level. Thus this methodology can bring at the end the possibility to have a full autonomous vehicle by leaving the command. However, if the confidence of the virtual driver decreases too much or the request from the driver is too important, a special sequence will be started to take back the driver in the control loop. Some standard procedures such as decreasing LED gauges or increasing torque on the steering wheel have shown good result by drivers.

4.3.4 Event-triggered intervention process

The intervention process will be started when the vehicle command is close to the boundaries of the safety envelope. Its main goal is to take the vehicle command away from these boundaries, and it can re-use the path transition algorithm to push the vehicle command in the direction of the corresponding optimum. If the virtual driver has been confident at 100%, it would stop the driver's command like the binary intervention. Furthermore if the virtual driver gets a lower confidence level, the limits of the safety envelope will only brake smoothly the driver's command variations. Thus the recovery plan from the virtual driver will be postponed depending on its confidence value and will have more or less power. On the following examples, the confidence of the virtual driver will be higher and higher until it takes the initiative to push back the vehicle command into the safety envelope (Fig. 4.12).

The braking capacity will be determined as an elongation capacity of the boundaries between the two points of the safety envelope. This choice has been influenced by the importance of the distance between the points. Actually, the closest are these points, the highest confident is the position of the points to be on the right position because those two points are working like a redundancy. Thus, far points may leave the possibility to the driver's command to deform the boundary such as in the same figure. Here again a fuzzy logic will be used to determine the stiffness λ of the boundaries. Let us re-define c the confidence of the virtual driver and $c^\star$ the confidence of the driver. The two

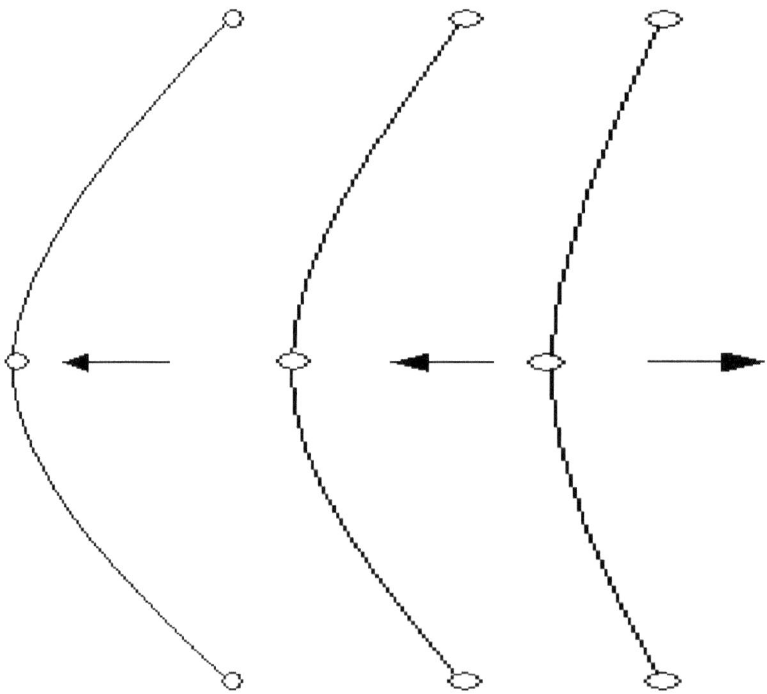

(a) Low braking capacity by low confidence level

(b) Braking capacity by confidence level equivalent to the driver's confidence level

(c) Push of the vehicle command by high confidence level

Fig. 4.12. Different levels of intervention from low confidence to highest confidence when the driver's command reaches the limit of the safety envelope most influencing phenomena can be set separately and multiplied together to define the stiffness (eq. 4.2):

- The confidence value c of the virtual driver. Its influence on the stiffness increases when it goes to 1 as shown in figure 4.13.
- The difference between the confidence values $c-c^{\star}$. The source having the most important confidence level will adapt the stiffness for its purpose: braking or not, as shown in the second part of the figure.

$$\lambda \approx \frac{1}{1-c} \cdot (c-c^{\star})^3 \tag{4.2}$$

4.3.5 Fusion of both processes

The fusion of both processes will happen when the driver's command is far from the optimum but not reaching the boundaries. Such cases can happen under two hypotheses. On one side, the deviation can be made voluntary by the driver. In this case, his/her confidence level will be high enough to have the virtual driver following the requirements. That is why it almost deals with the intervention process. This intervention has to be delayed as long as possible for a really confident driver and the substitution process has to be disabled. In the figure below, the variation of the influences will grow for the intervention process but be delayed and the substitution process influence will be set close to zero. On the other side, the unwilling deviation can be made by a not-aware driver. In this second case, the virtual driver has to be clearly a substitute to the driver. Thus the variation of the influences will be at the opposite of the other case (Fig. 4.14).

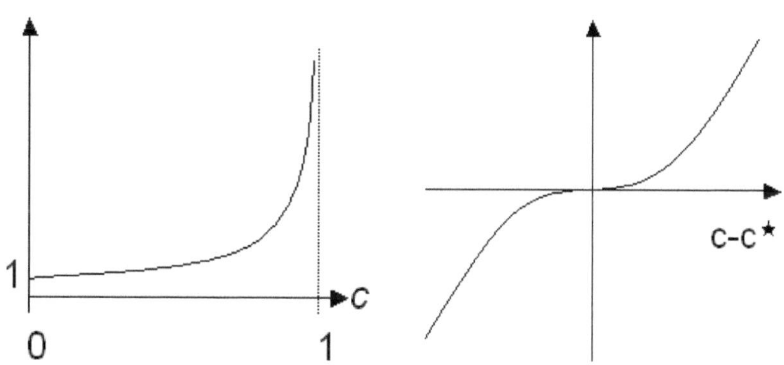

(a) Influence of the virtual driver confidence c

(b) Influence of $c-c^{\star}$

Fig. 4.13. Fuzzy rules for the intervention level

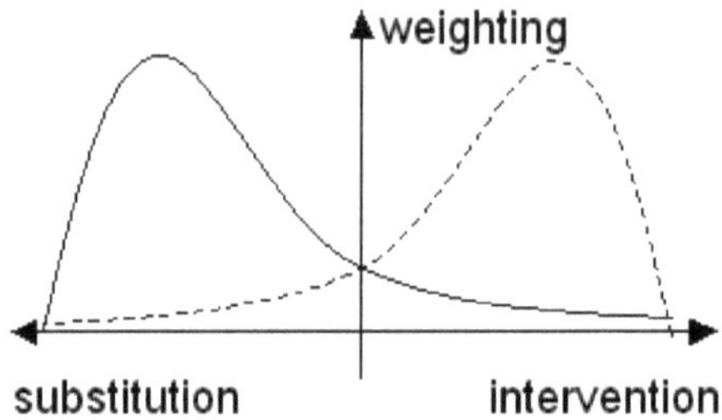

Fig. 4.14. Varying influences of the both processes depending on the confidence level of the driver

4.4 Results and analysis

The first cooperation algorithm presented in this chapter has been a binary intervention. As soon as the driver's command reaches the limits of the safety envelope, the decision control will realize a transition for the vehicle command from the current dangerous driver's command to the optimum of the corresponding maneuver. Such a kind of intervention, who is only a two-dimensional-improvement of the braking assistant systems such as the emergency brake from DaimlerChrysler, is making a strong hypothesis. Actually, the virtual driver is supposed to get at each time all the information about its environment and about the vehicle, which leads to a perfect determination of the limits of the safety envelope.

As such level of perfection cannot be reached, two main restrictions have been added. First, the position of the limits is only rough advices to provide a gradual intervention. Secondly, the level of intervention from the virtual driver side is depending on the confidence values of both driver and virtual driver. Hence the risk of wrong intervention is lowered by looking whether the system can intervene or not. However, the gradual level of intervention when the driver's command is close to the limits of the safety envelope, and this phenomenon can be seen by the driver as restriction of his/her freedom. That is why the driver feedback has to be drastically more influent to let the driver understanding how good he/she can be supported. This new way of feedback will be described in the next chapter.

5

Feedback management for the driver and the virtual driver

Abstract. The huge importance of the feedback to support the driver has been shown previously over the last chapters. By giving the safety reserve of the vehicle, the driver can manage dangerous situation better and he/she can avoid reaching the limits of the safety envelope. Furthermore each intervention performed by the virtual driver on the vehicle has to be explained to the driver for two reasons. First of all, the driver should understand what is happening and above all why the intervention control takes the initiative to let him/her improve his/her driving skills.

This chapter will start with the description of the analogy between the feedback management and a Delphi method. After that the feedback to send to each actor in the vehicle will be described. First, the feedback to the driver will be extended to integrate information about command and executive levels. Secondly, a feedback to the virtual driver will be created in order to check indirectly its working hypothesis. With such a concept both actors will have the possibility to verify their cognitions about their environments and to testify their confidence levels.

5.1 Analogy to the delphi method

The methodology of the Delphi method, as explained with more details in [78], for example, is based on the consultancy of several actors to resolve a main problem. Each actor will provide a solution with a priori weight of his/her knowledge relative to the other actors. Then, all solutions are combined depending on those weights. The next step is to send back the resulting solution to each actor. They will have the possibility to correct or not their proposals with this information. Finally the definitive solution will be extracted after this last call.

Now the analogy between this method and the intervention process will be used to find some missing links within the data flow between driver, virtual driver, intervention control and executive level. Let us take the two actors driver and virtual driver, who try to solve the following problem: what is the best vehicle command to realize safely the maneuver chosen by the driver. Both actors deliver their solutions by means of the driver's command and the corresponding optimum. Indirectly a confidence value is generated by the driver and virtual driver, based on the drowsiness monitoring and the self-estimation algorithm within the virtual driver.

After the fusion of these possible solutions, a command will be sent to the executive level as the best-known vehicle command. In parallel, a feedback is sent to the driver by giving the limits defined by the virtual driver. Thus, such action can be performed only if the virtual driver has got a full confidence. In this case, its solution is taken de facto as final solution when an intervention is required. Hence, only a feedback to the driver is required as the virtual driver does not need any advices. Moreover the driver's command is supposed to be used to define the maneuver, but the execution level will not take it into consideration at all: the virtual driver's optimum will be realized as given.

As already explained in the preceding chapter, such full confidence is not possible. Thus, feedbacks have to be sent to both driver and virtual driver as hints about the acceptance of their solutions. The driver will get a feedback about the tactic level, which corresponds to the choice of the maneuver, and about the executive level, which corresponds to the quality of the realization of this maneuver. On the other side, the virtual driver will only get a feedback about the executive level. As only the driver is responsible for the tactic level, the virtual driver will not receive any feedback about this. Its knowledge will be used anyway to define the feedback to the driver.

5.2 Detection of partial and full conflict situations

A feedback is sent continuously to the driver and to the virtual driver in order to support them to improve their commands. Depending on the difference between their commands and the feedback, two different behavior models can be used. During a partial conflict, the intervention control sees a gap between the driver's command and the optimum from the virtual driver. Both, driver and virtual driver have chosen the same maneuver but there is a small difference either in the perception or in the skill to perform the maneuver. This kind of conflict is not dangerous because both have chosen the same maneuver and realize it with minimal differences.

However, when the distance between the command and the feedback is growing too much, a full conflict has to be handled. In figure 5.1, the virtual driver did not observe the temporary markings. Thus, the virtual driver would not like to go to the right, which is not compatible with the steering request to the right from the driver. Hence, the virtual driver with its high confidence will intervene dangerously on the vehicle command. However, with a feedback about this conflict, the image processing, for example, will have the possibility to verify if another lane marking corresponds to the driver's command. When

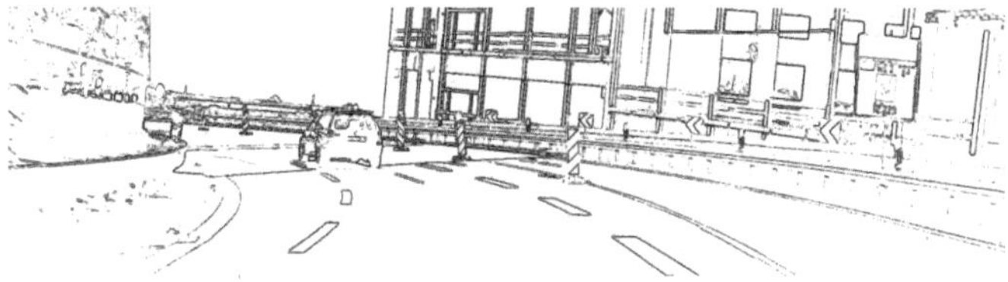

Fig. 5.1. Example of local calibration

a positive result happens, the virtual driver will automatically regenerate the safety envelope and the maneuvers' characteristics.

In the case of full conflict, the driver's command tends toward the boundaries of the safety envelope. This conflict is based on a different choice of maneuver and may come from a lack of information or misuses of the driving rules. If a full conflict is detected, an analysis will start by extrapolating where the driver's command will cross the boundaries of the safety envelope. For each type of intersection, there are many possible hypotheses to check as reported in figure 5.2.

If the intersection is at the limits of the maximal acceleration, there will be four different hypotheses to check: an object ahead, maximal vehicle capacity, speed limitation and road curvature. If the intersection is at a limit of the minimal acceleration, there will be only two possibilities: other vehicle or the vehicle capacity. For an intersection at the limitation of the steering range, the first factor to check will be the width of the current lane; the next steps

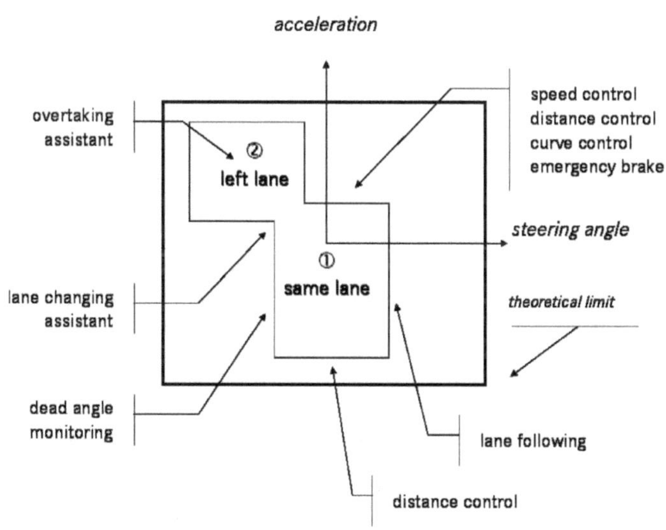

Fig. 5.2. Hypothesis on the zone of conflicts

are the existence of another lane, the type of the lane marking, the presence of other vehicles and the maximal capacities.

The data fusion triggers the sensors to check the viability of the environment of the virtual driver one after the other. For example, the object detection can be verified by the radars if they did not use a too low threshold to determine if a point is a ghost or not. The image processing can verify the position of the marks and their type by giving the possible position.

At the same time the decision control system sends feedback to the driver to verify his/her perception. For example, a warning could be shown on a re-oriented mirror for the dead angle monitoring. Moreover a sensitive feedback can be performed with a resistive torque describing the direction of the driving optimum.

5.3 Feedback to the driver

A feedback is defined as the information for a human about his/her behavior or performance [130]. Moreover most of the humans like to get advices, since it helps them to improve their skills [39]. Studies on passenger/observer feedback have shown that individual feedback is an effective tool for positive behavior modification [61]. Other studies such as [138] suggest that the introduction of vehicle data recorders alone, without giving feedback to drivers, might increase safety in transport operations and reduce traffic accidents: people who are aware of being observed tend to modify their behavior.

5.3.1 Generation of a feedback for the driver

The feedback generated here is relative only to the command of the vehicle depending on the vehicle state and the dangerousness of the command. The other elements such as the tank level are not provided through this function but directly linked to the dashboard.

At each moment, the driver can get a feedback on both tactic and operative levels. The difficulty here is to ensure a good fusion of those feedbacks, which is evident enough to avoid bad interpretation from the driver. For each level, the feedback to perform will be generated and then they will be fused together.

Feedback from the tactic level

The driver requires information about the maneuver or the way to perform safely a transition to another maneuver. This information only depends on the safety envelope. Several levels of feedback can be used from pure visual feedback to feedback with all channels prior to intervention. Thus thresholds for the acceleration and the steering angle will depend also on the cooperation grade as shown in figure 5.3.

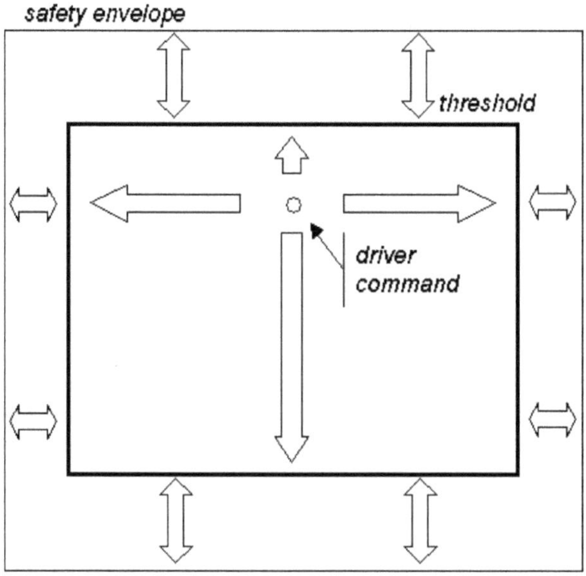

Fig. 5.3. Position of the thresholds

Nonetheless, the identification of the maneuver can be checked indirectly here by looking at the acceptance of the driver to the feedback. Thus the feedback relative to the optimum choice will have to be one of the most important parameters in order to avoid having it polluted with other information.

On one hand the thresholds will come sooner when the assistant system is really confident. On the other hand, a fit driver, which is aware of the sources of danger, will have delayed thresholds. The position of these thresholds will use directly the results of the fusion done before.

Feedback from the operative level

At the operative level, the driver has to be aware first of the level of assistance he can receive. And secondly the driver needs also a feedback about the vehicle state to close his/her control loop. Thus, the feedback will be based on the distance to the optimum and on the position of the vehicle command, which is the result of the cooperation. For practical reason, sidestick or steering wheel can be configured with a passive torque, which can get its neutral position on this optimum. With this mechanical resistive torque, the driver will be able to correct his/her command in the right way.

On top of that, the vehicle state will be provided. The driver gets the difference between his/her command and the vehicle state to manage the control of the vehicle in the best way. Furthermore this feedback needs to be disabled as soon as the executive level shifts a gear. Otherwise the feedback will be shortly sliding because of the null acceleration and may disturb the driver.

If there is no optimum defined for the maneuver, a virtual optimum will be used to help the driver. Actually an optimum may not be defined if there is trouble with the longitudinal control. In this case the acceleration of the optimum will be the same as the optimum of the driver's command. Furthermore for the lateral control there is always a possibility by giving the middle of the local safety envelope.

Framework for the fusion of different elements

The feedback $\mathcal{F}$ is implemented along the lines of [108], where a virtual force framework for vehicle control is developed which allows superposition of virtual forces to be applied to the vehicle that can represent control inputs from tactic and operative levels. The resulting feedback is defined as the summation of all the feedbacks.

In practice, a limited potential field $\mathcal{PF}$ (eqn. 5.1) is applied to each feedback and for each direction (acceleration and steering angle), by looking at the distances $(\Delta_\gamma, \Delta_\theta)$ between the driver's command and the corresponding maximal and minimal accelerations and steering angles, the optimum and the vehicle state. The potential field has to be defined correctly for each feedback with different amplitude parameters A and B, and what the limitation of the feedback amplitude is.

$$\mathcal{PF}(d) = \begin{cases} \frac{A}{\Delta_\gamma^2} \\ \frac{B}{\Delta_\theta^2} \end{cases} \tag{5.1}$$

$$\begin{aligned} \mathcal{F} = & \mathcal{PF}_{optimum} \\ & + \mathcal{PF}_{limits,acceleration} + \mathcal{PF}_{limits,steering} \\ & + \mathcal{PF}^{-1}_{vehicle,state} \end{aligned}$$

In normal cases, the driver's command will stay wide away from the safety envelope limits. The effect of the potential field for the optimum has to be limited in order to have it lowered by the effects of the safety limits when they are really close to the driver's command. Moreover the effect of the vehicle state also has to be taken into account. Indeed, when the distance to the vehicle state is important, the force has to attract the feedback in order to see the driver reducing this difference between command and vehicle sliding (figure 5.4).

5.3.2 Different used channels

The driver gets roughly three channels to get advices as shown in the picture of the mockup (fig. 5.5). This mockup has been installed during the SPARC project with the following elements described later:

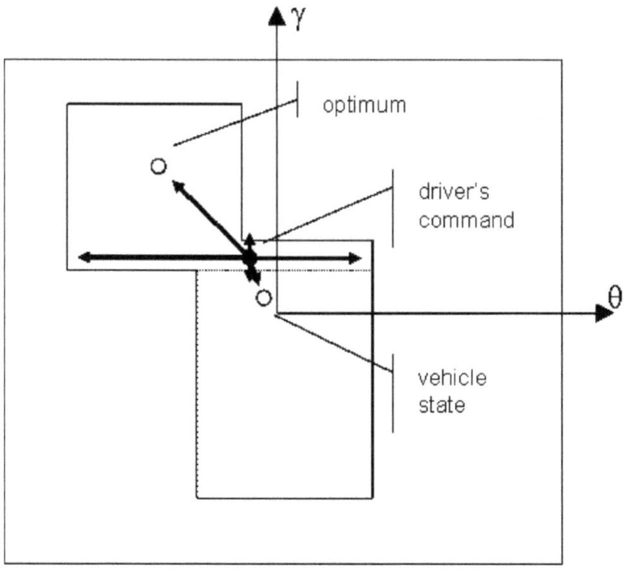

Fig. 5.4. Advices to give to the driver

- Visual channel (channel 1)
  - Display for standard information
  - LED displays for lateral control and longitudinal danger
- Auditive channel (channel 2)
  - Lateral control by emulation of tires noise over a lane marking
- Tactile channel (channel 3)
  - Reflexive correction

The tactile channel will be used with the integration of two advices:

- Stiffness of the sidestick depending on the distance to the optimum. The optimum is directly used by the sidestick as neutral position and will tend to go to this position. Therefore the driver will feel this difference even if it is not important enough to move directly the sidestick.
- Vibration in case of emergency. As soon as the command is closed to the limits, omni-directional vibrations will be performed. Their amplitude will depend on the distance to the limit in order to get advices about the dangerousness.

5.3.3 Feedback dispatching

Depending on the dangerousness and the driver's fitness, a different channel will be used. Standard advices about the vehicle and non-dangerous situation will be displayed through the visual channel. Auditive channel will be used

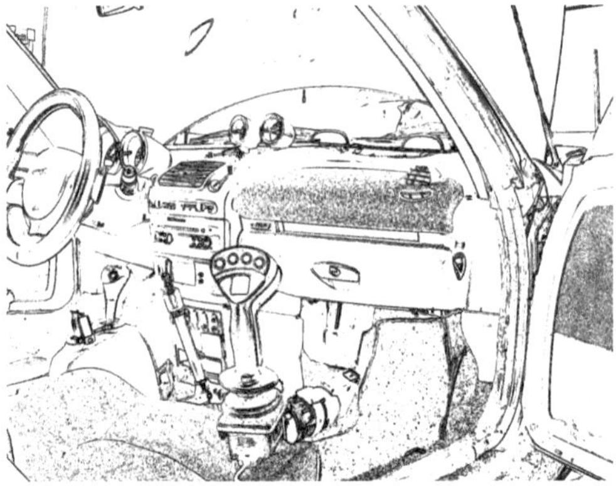

(a) LED displays and sidestick integrated into a smart roadster for acceptance tests

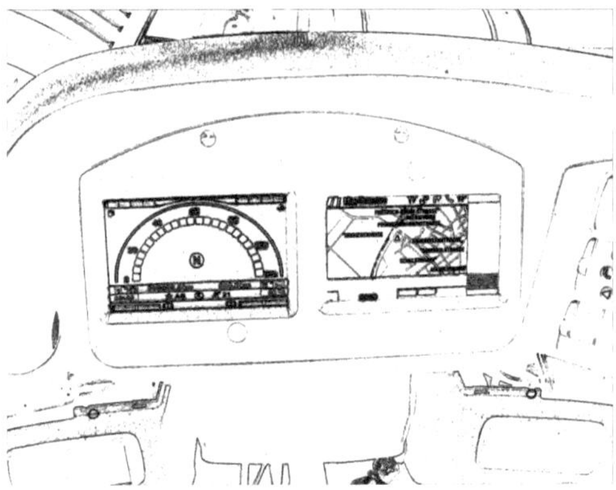

(b) Modified display from an Actros to show the sources of danger on the screen

Fig. 5.5. Integration of the different feedback channels within a passenger car and a heavy good vehicle

only for the lateral correction because the driver gets a worse spatial representation capacity with the auditive channel than with the visual channel.

The choice of the active channels and their levels of activity is growing with the increase of the feedback level. Two different methodologies are currently under investigation, for example, the feedback management by Huang [60], visual attention by Gibson [47] or Theeuwes [126], auditory attention by Hansen [56] or by Mondor [90]. The first case (depicted in figure 5.6) enables a new channel once the current one has reached its maximum activity level. The second case (depicted in fig. 5.6(b)) enables a new channel before the current ones have reached their maximum levels by using a constant ac-

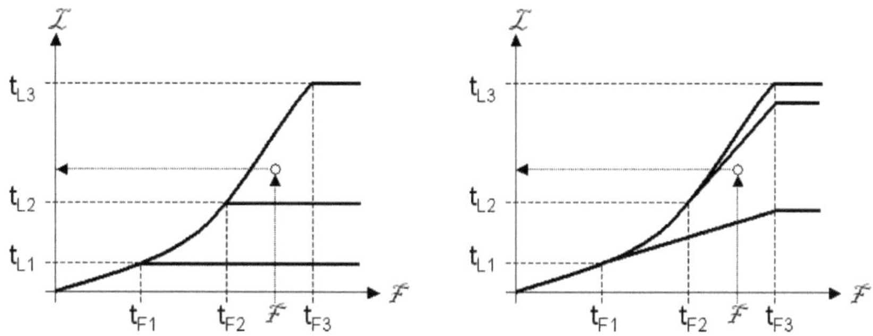

(a) Sequential use of the feedback channels

(b) Additive use of the feedback channels

Fig. 5.6. Possible choice of the feedback dispatching and the influences on the level of feedback for the different channels

tivity slope. Hence a transition between the main channels can be performed over the increase of dangerousness.

Depending on the feedback level $\mathcal{F}$ and the dispatching concept, the active channels will differ. In the first case, $\mathcal{F}$ is included within the third channel range $(t_{F2} - t_{F3})$, thus both visual and auditive channels are fully active and the residual difference $(\mathcal{F} - t_{F2})$ is used to define the stiffness of the sidestick. However, in the second case, only the auditive and visual channels will be used. In practice the second strategy (fig. 5.6(b)) with low slope has received a more positive acceptance from testing drivers on simulation platform. Most of them have appreciated that all the active channels get more activity among the dangerousness without having any rupture of the channels activities. Otherwise, once a channel has reached its maximal activity level, the driver would not like to perceive it anymore as it does not bring any additional information anymore.

Two possibilities have also tested for the slope definition: parallel slopes and slopes defined as local tangent. In the first case, the drivers have appreciated the same variations on all channels which do not favor any channel to catch the driver attention abruptly. However, in the second case, the test drivers have appreciated the continuous variation of the channel use (in contrary to the first case where abrupt variations appear at the thresholds).

The qualitative tests performed with a vehicle simulator have shown some interesting results. First of all the use of the safety envelope combined with the feedback dispatching has reduced efficiently the complexity of the interface with the driver, which helps him to react faster to the feedback with a lower workload. The next step will be a self-adaptation of the interfaces in order to optimize the feedback dispatching in terms of thresholds, transition models, etc. Such generic learning feedback management requires a driver model to adapt continuously the strategy depending on the driver state (e.g., activity [82], frustration [86]) and the previous decisions.

5.4 Feedback to the virtual driver

In the preceding section, the information for the driver was clearly determined but the main problem was still to find the adequate way to bring it to the driver. Unfortunately, indirectly getting the information now from the driver will be more difficult. However, looking anyway at the driver with different methods will help the virtual driver to get some advices, even if this kind of feedback cannot be considered as safe enough. In practice its influence will be lowered down to second-order information with safety restrictions.

Let us have a look once again at the driver model. Three serial functions are relevant here to describe the information flow from the environment down to the vehicle command (figure 5.7):

- Monitoring the surrounding environment
- Choosing the adequate maneuver
- Realizing the maneuver

Getting information about the environment is possible by two ways. A first method is based on monitoring the direction of a driver's eyes: looking where the driver is looking. As the driver gets most of the information over the visual channel, an eyes monitoring with internal camera like in [69] is an easy way to grab information about possible regions of interest. As soon as the driver's eyes temporize shortly in a direction, road-users, traffic signs, etc., can be expected in this region of interest (ROI). This direct method will not be used here because the acceptance of internal camera by the drivers is unfortunately quite low (same problem as for direct drowsiness monitoring, see Chapter IV.3). The second method is an analysis of the driver's command. Depending on the driver's correction, the influence of new road-users can be determined with some probabilities. This approach will be focused on now for the lane detection and the position of other road-users.

5.4.1 Check of conflict due to lane detection

Two phenomena can be extracted during the driver's command analysis for the lateral control. On one side, sudden steering modifications can happen normally in two cases. Either the driver was falling asleep and he/she woke

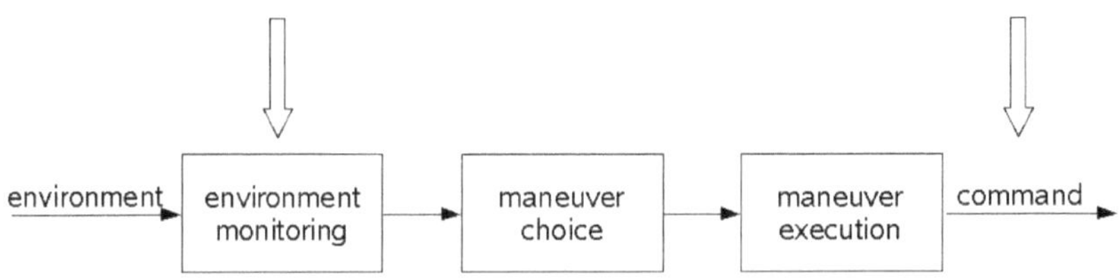

Fig. 5.7. Driver cognition model used to grab information about the environment

up frightened, or the driver has made a jerk. When the driver is drowsy, his/her confidence value will also be lowered and in this way its influence will be lowered. Next, the jerk maneuvers are cortex reflexes to avoid any collisions.[1] Unfortunately these maneuvers are performed in open loop as the couple muscle–cortex has no capacity to manage any control loop [54]. Thus, contrary to a controlled maneuver managed from the driver's brain, such jerks are always more or less equivalent without really taking into account the source of danger. Hence such actions will not provide any advices to define the position of the road-users.

On the other side, steering corrections can bring more information as they are controlled actions. The main idea as shown in the previous example (fig. 5.1) consists of looking whether the driver accepts or not the feedback by changing his/her command. As soon as the driver makes an opposition to the feedback during more than 1 s (threshold found empirically), the image processing will look at another possible lane in this direction. In fig. 5.8, the image processing did not detect the creation of the left lane. However, due to the advices from the driver resistance on the steering wheel, a ray tracing (black lines) has been started on the left side and it has found the missing lane to accept the driver maneuver.

Fig. 5.8. Ray tracing on the left initiated by a conflict with the driver in order to verify wheter a lane exists on the left

[1] These reflexes are actually like taking his/her hand away from an hotplate The muscle command is directly generated within the cortex and never goes up to the brain.

5.4.2 Check of conflict due to road-user detection

Normally monitoring the other road-users shall be more stable than the lane detection because it is possible to fuse data coming from redundant sensors (e.g., camera, radars, laser scanner). Hence it is possible to compensate the errors coming from one sensor with corresponding results from other sensors. Moreover an object tracking over the time can improve the accuracy of the positions. Furthermore in some cases, it will be anyway interesting to give regions of interest to the sensors depending on the driver's behavior.

In practice, the conflicts between driver and virtual driver due to other road-users can happen with two patterns. On one hand, the driver may start a forbidden maneuver. In this case, the hypotheses disabling the maneuver have to be checked by the corresponding agents. Actually, most of the time, the presence of road-users has to be validated once again. On the other hand, the driver may refuse the help as it corresponds to a wrongly understood maneuver. For example, the driver has seen other road-users (e.g., pedestrian or vehicle ahead) that the sensors did not detect.

Radar triggering

An object detection set using two short-range radars have been developed with the possibility to trigger regions of interest. Two identical 24 GHz radars have been integrated behind the front bumper of a passenger car and each radar sends every 10 ms a list of 10 reflections on a CAN bus[2] with the attributes of distance, bearing, relative speed (over Doppler effect) and reflection level. The analysis of these data is performed in three steps (see fig. 5.9): deleting the reflections less than 70 dB, clustering the reflections to determine the objects, tracking the clusters to improve their information. Errors can arise on four locations: first the message may not be sent over the CAN bus,[3] secondly the

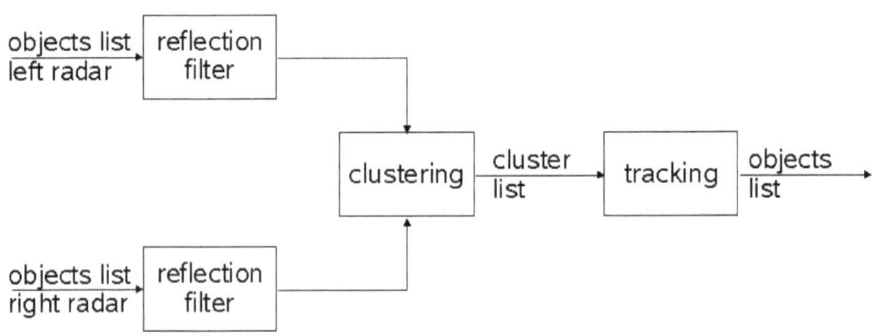

Fig. 5.9. Dataflow of the radars-based object detection

[2] Controller area network.
[3] Not time deterministic, can be avoided by using FlexRay.

reflection may have been rejected as it is too low, thirdly the clusters may not have been created correctly and finally the tracking may have detected inconsistencies.

When the region of interest is defined, the first hypothesis to verify is the existence of ghost objects. A ghost object is the result of a wrong interpretation which leads to the creation of a non-existing road-user or the non-creation of a road-user as it is supposed to be a noisy result. Hence low reflections will be searched close to the region of interest. In normal cases, reflections less than 70 dB are due to echoes such as trees or houses. However, time to time some reflections corresponding to pedestrians can be wrongly interpreted as peoples have an important wave-absorption coefficient.

Figure 5.10 represents measures taken at a crossing section during 10 s. The reflections with a distance less than 2.5 m correspond to pedestrians on a pedestrian crossing. Furthermore the reflections by 3 m correspond to vehicles with the exception of the reflections at 9 m. These last reflections are only noise and show the 1° bearing step of the radars. The intensity of the reflections from the pedestrians is most of the time as low as noisy

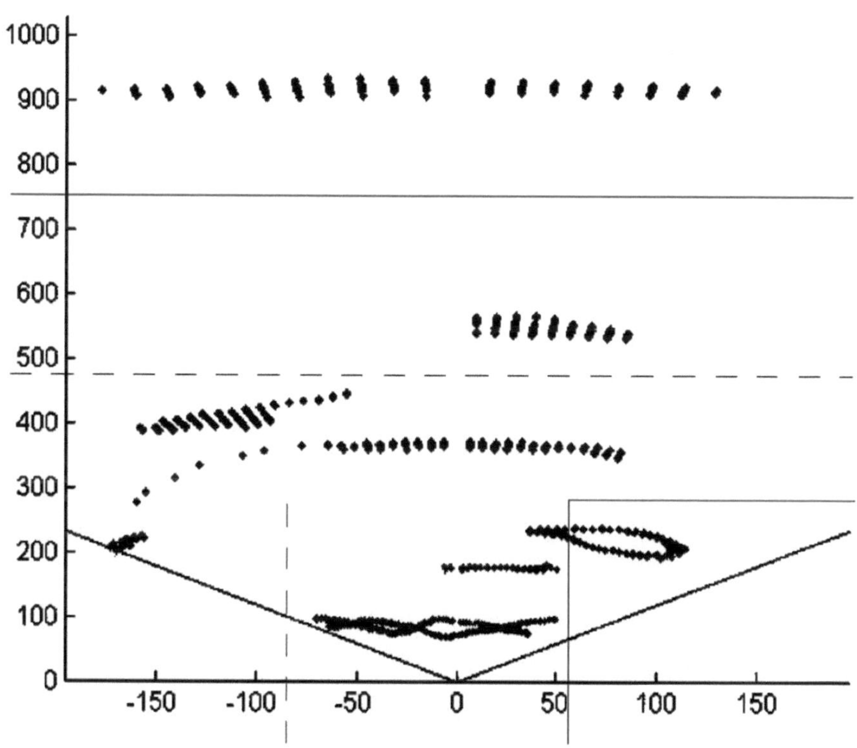

Fig. 5.10. Level of reflections from pedestrians at a crossing section

reflections when the pedestrians are close to the limits of the radars' range. Thus, the pedestrian detection would be postponed until the reflection reaches the required threshold. Hence the use of an external trigger can help to detect faster the pedestrians and to save some precious milliseconds.

The second hypothesis deals with the quality of the clustering. The method used here is similar to the clustering of optimums presented in Chapter III.1, Section 1.2.3. Prior to the online clustering, an off-line calibration of the sensors has been done with a standard metallic reflector. Figure 5.11 presents the variances of the measures of position for this reflector. After that, the clustering method will use these variances to determine the uncertainty range of each reflection. Results of such clustering are presented in figure 5.11 where two objects have been detected. In practice mistakes may be done when a

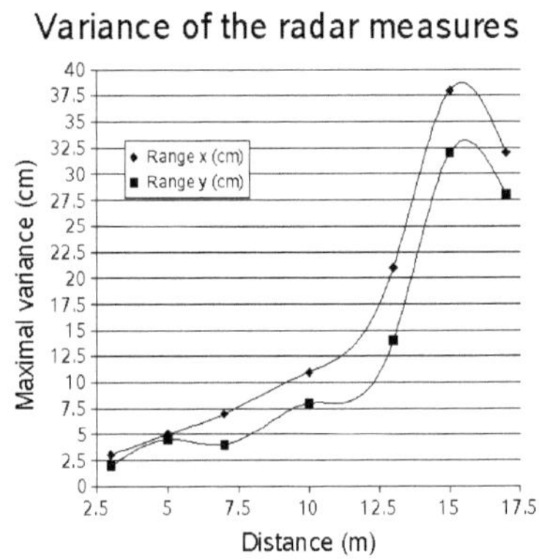

(a) Longitudinal and lateral variances of the radar measure depending on the object distance

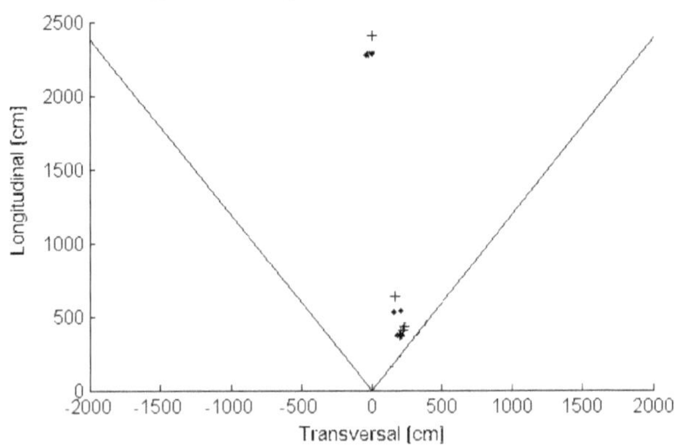

(b) Creation of clusters with two parallel radars

Fig. 5.11. Use of radar variances to determine object clusters

(a) Definition of the mean value for the background subtraction

(b) Detection of the vehicle with the shadows

Fig. 5.12. Vehicle detection by use of a background subtraction to detect abrupt changes group of pedestrians may be seen as one vehicle because a vehicle can be seen as multiple reflections too. However, such errors of clustering do not weaken the results of the detection as the presence of a road-user is enough to adapt the vehicle's maneuver possibilities, even if the type of user is not correct.

The last hypothesis is linked to the object tracking. This tracking based on a linear Kalman filter [21] uses a dynamical model of the tracked object. Here different models can be applied depending on the possible object to track: pedestrian, passenger car (high dynamics) or truck (lower dynamics). When a region of interest coincides with an object tracking, the Kalman filter will not have the possibility to disable the track even if the dynamics is not compatible with any model.

Camera triggering

Several methods exist to detect object on a single daylight image as presented, for example, in [55]. Among all these methods, three standard methods have been used here in parallel (fig. 5.13): background subtraction/color-based extraction, patterns extraction, image flow variations. Each method provides a list of detected objects, which are combined to improve the stability of the results. The use of external trigger with possible regions of interest is described in this section only for the background subtraction-based method. However, the concept is generalized to all methods as they are all based on two main parts: definition of the regions of interest, verification/tracking of these regions of interest.

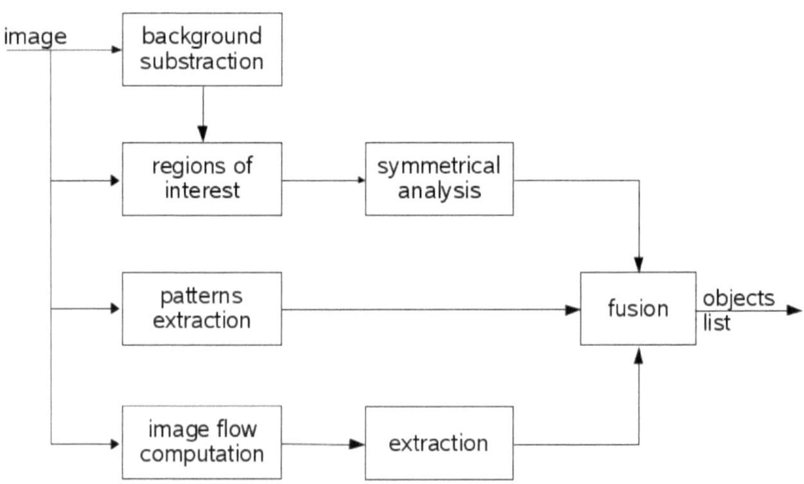

Fig. 5.13. Image processing dataflow

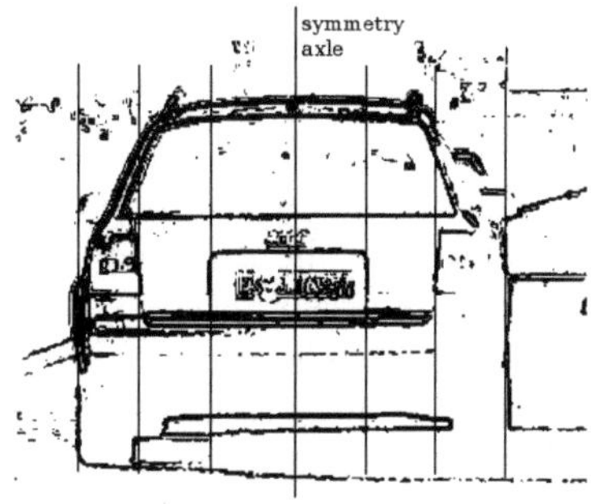

(a) Example of passenger car and the vertical vertices to be detected

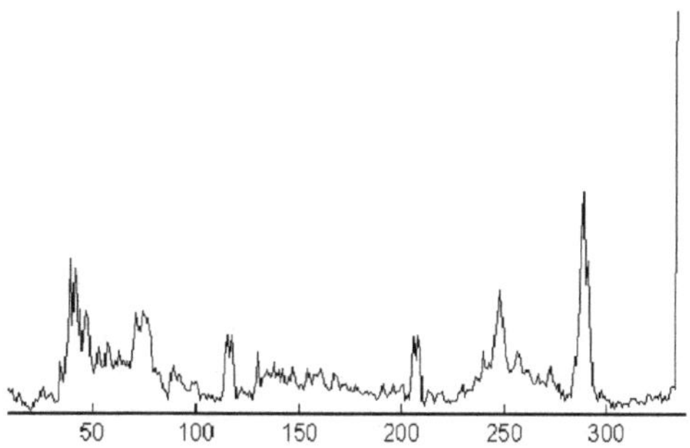

(b) Corresponding histogram of relevant pixels

Fig. 5.14. Computation of the histogram of relevant vertical pixels to find symmetries

The first step is the extraction of the background in the picture. In normal cases, the background can be defined almost as a constant gray level with some variations. A standard point ahead the vehicle, which may not represent any road-user (as in the figure 5.12(a)), is used with a small range around it to define the background main gray intensity as the mean value.

After that the full picture under the vanishing point is scanned to remove the pixels within this luminance range. Then an erosion of the picture is performed to remove the single noisy pixels. The other road-users will still be present on the picture because their own gray level or their shadows will have edges that cannot correspond to the background (examples in fig. 5.12(b)). Finally, the regions of interest have to be matched on these pixels to form rectangles. Those rectangles will be verified during the second step.

The possibility to provide here additional regions of interest will enable the image processing to verify other assumptions. The second step takes each regions of interest and verifies the presence of another road-user within this sub-picture. Here again, different methods exist and may be used in parallel to extend the reliability of the object detection. Patterns can be matched on the regions of interest to verify the forms and to extract the type of road-user at the same time. Actually the background subtraction is in this case just a way to determine quickly and roughly the sub-pictures instead of having multiple parallel searches. However, this method requires an important amount of computation time and capacity while it matches all the possible patterns to determine their probabilities of acceptance.

Thus, another method will be used here, which is faster but cannot provide the type of road-user at the same time. On one side vehicles are built instinctively symmetrical for esthetical and aerodynamic purposes. On the other side, natural elements are seldom symmetrical because of their stochastic aspects. Mathematically speaking, the symmetry is described as the corresponding rate between left and right sides of the symmetry axle. However, it is not possible to make a direct comparison of the pixels due to differences of luminance and vehicle orientation. Hence an analogy of the vertical vertices is used as solution in practice because it is more robust to these phenomena. First of all, a derivation of the sub-picture is performed by using a Sobel mask. After that a threshold defining the significant edges is extracted from the corresponding histogram by finding the 70% of the energy within the histogram.

Once the region of interest has been binarized by comparing the derivate with the threshold, the histogram of the amount of relevant pixels per column is calculated. Figure 5.14(b) corresponds to this histogram for the picture in fig. 5.14(a). The peaks corresponds to the position of important vertical vertices. Finally the cross-correlation of this histogram is calculated to check if a symmetry axle exists with a good rate ρ (in practice ρ is supposed to be greater than 0.8 as on this example and shown in figure 5.15).

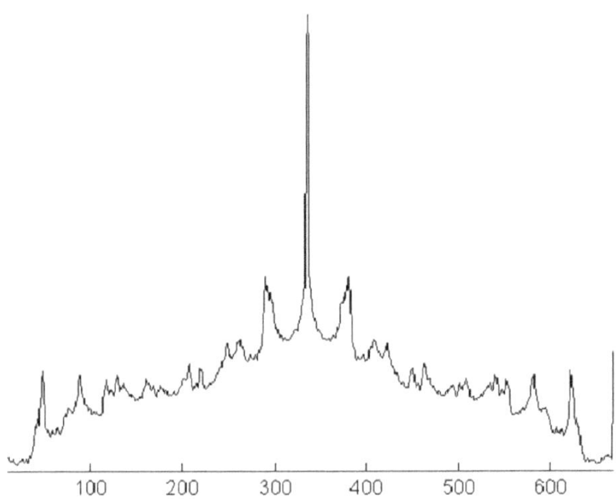

Fig. 5.15. Histogram cross-correlation with symmetry axle at the pixel index 310

5.5 Critics on the new feedback extensions

This chapter has presented the driving task as a Delphi method where driver and virtual driver bring their knowledge to optimize the vehicle command. As no actor can bring the full solution, the equilibrium needs to be found by sending a feedback to both actors. Hence, they will have the possibility to take into account new hypotheses or to increase their confidences.

The information for the driver can be collected without important difficulties but the main problem is to bring them adequately as the driver's acceptance of feedback is continuously changing. Hence a framework has been built where generic parameters can be adapted to each driver.

On the other side, the problem is more critical as for one driver's action, several hypotheses have to be verified at the same time. Even if it is possible to sort those hypotheses depending on their probabilities of occurrence, the verification time will be huge (about 1 s) in order to avoid false acceptances. That is why feedback to the virtual driver can only be used to correct smoothly its environment perception. Actually it is possible to react really faster but with a much higher risk of false acceptances.

Part V

Discussion on the proposed concept

6

Concept summary and overview of the functionalities

6.1 Needs to help the driver in his/her task

From a statistical point of view, even if the driver avoids more than 99.99% of the accidents, he/she is still responsible for the accidents in more than 95% due to insufficient environment perception or wrong response. Thus if one wants to make the road safer, it is mandatory to focus on the driver in order to improve the reliability of his/her command. Hence, several methods are under investigation. On one side artificial drivers are integrated within the vehicle to create autonomous vehicles. As the driver has been taken out of vehicle control loop, the vehicle should have been safer. However, such concepts still have high difficulties in urban environment because of its wide complexity. Moreover the adaptation capacity of such devices is not as good as by an experimented driver. On the other side, the main hypothesis is to keep the driver in the control loop as no algorithm may be as flexible and preventive as the driver himself/herself. Hence devices are required to support the driver in his/her task of vehicle controller: feedback, drowsiness monitoring, etc.

However, no one of those two solutions is sufficient due to heavy drawbacks. Indeed, they can be seen as complement because it is interesting to have the possibility to use the better of those two possibilities depending on the situation. Hence, a heterogeneous redundancy of the vehicle controller can compensate one failing actor by relying on the other one. This new concept has been described here and the whole vehicle concept has been re-worked in order to integrate this novelty properly.

6.2 New vehicle architecture concept

A new vehicle concept has been conceived here to integrate both concepts at once. It is mainly split into five modules which enable the clear integration of functionalities with complexity reduction on top of that. Indeed, up to now commercialized vehicles are integrating their devices with a vertical layering.

For each functionality and dynamics control a couple sensor(s), intelligence, actuator is defined. However, this architecture has reached its integration limits for several reasons. First of all, the prioritization of the actions on the actuators has to be validated as soon as new functionalities are added. Secondly the required communication bandwidth is no more cost effective. Indeed the synchronization of the longitudinal and lateral dynamics is hardly performed on decentralized architecture. Thus a horizontal layering concept, which is already used in robotic and in airplane fields, has been introduced here. The main idea is to concentrate all the functionalities dealing with the same kind of control on the vehicle to encapsulate them and to minimize the rest of the communications.

Two main layers have been defined. First of all, the command level has been defined where the driver defines his/her driving strategy and gives a command to the vehicle. Secondly, the vehicle realizes the command that it gets from the command level and it sends a feedback to the driver about its capacity. The integration of the assistant systems as a virtual driver on the command level allows the control loop to be redundant. Thus, an intervention control has been required to fuse their commands and transfer the safe resulting command to the execution level. With this kind of architecture, the assistant systems have no direct connections to the execution level and they can make their prioritization or synchronization locally without any influence on the rest of the vehicle. On top of that, a human interface is required as wrapper for the driver to acquire his/her command and to give him/her advices back. Finally the last two of the five modules are integrated. They correspond to the comfort modules, which include all non-driving relevant functionalities such as the radio or the air climate, and the wiring harness module with its connections to the body controllers. With this new approach the communication between modules are defined without taking into account the level of integration in each module. Thus, the whole vehicle architecture is validated independently from what is integrated within the modules. Moreover each module can be upgraded later without any influence on the other modules. Hence, the risks and the complexity can be decreased.

6.3 Creation of an extended executive level

The main change in the executive level concerns the synchronization of the actuators. Up to now, this synchronization has been done by the driver, which has to manage one physical input per actuator (e.g., pedals, gear box, clutch, steering wheel) or with some partial automation such as automatic gear shift. Hence the intervention on the driver's command can only be a correction as, for example, with an active front steering. However, such reactive corrections cannot provide a stable vehicle control as both actors are still competing: a correction noise will ever stay and make the general vehicle handling uncomfortable. Hence, the choice made here has been to relocate the vehicle control on a central command which performs only one single command coming from the command level. That means, the driver provides only a command depending on his/her strategy as acceleration and steering angle.

This central element working first as gateway for the command level receives continuously a vehicle command to perform and a command window, which defines the maximal acceptance of the vehicle state. Depending on the vehicle state, its global stability control determines the slip commands for each wheels. This global control requires a geometrical model of the vehicle and the trailer (dolly, semi or full trailer), the position of the center of gravity and the body masses, which are indirectly determined with a vehicle observer. To perform this local slip command, energy is required from the energy reserves: combustion engine, electrical engine and batteries or capacitors. The energy management triggers each source directly to provide sufficient energy level to bring the required energy down to the wheels and to ensure a sufficient level of electricity reserve in the batteries. Here additional information such as the road slope or the type of environment is used to optimize the strategy (e.g., on descending road the electrical engine can be used anyway as it will charge the battery in parallel to its action, or exhaust gaze after-treatment cannot be used in city as the exhaust pipe cannot reach the required temperature).

After that the different energies are added together with a torque vectoring to provide the right energy level to each wheel. First a clutch distributes the energy on the different front and rear axles. Then a differential splits the incoming energy to the wheels of the axle. Finally on each wheel a local control performs the slip command by using a wheel dynamics model. At the wheel level, the different elements (damping control, steering actuator, brakes, and intelligent tire) can be encapsulated together to create a generic electrical corner actuator, which can be used indifferently on each front and rear axles.

If one wants to have the virtual driver highly accurate, it will have to feel the vehicle capacity like the driver does. Hence, the executive level also has to provide a real-time dynamics model to the command level in order to help it to optimize the computation of the vehicle's safety envelope, which requires determining extreme maneuvers with high accuracy. Hence an identification of the vehicle dynamics is performed at each time to determine the latency time, the dynamics rate and the maximal capacity for acceleration, braking and steering. Such a model corresponds to a second-order transfer function. The introduction of a model extension to describe the temporary state during a gear shift did not receive a great resonance at the command level as it made the algorithm too complex for any real-time usage. Hence this extension has been suppressed, even if a second-order transfer function has not described correctly the vehicle dynamics for an infinite time as accelerations are always tending anyway to zero at one moment (by reaching the maximal speed or null speed).

6.4 Integration of a virtual driver

The main goal of the creation of a virtual driver is to integrate all the assistant systems together in order to have a scalable module which makes the rest of the vehicle architecture independent of its content. Up to now these functions are used only for comfort purpose. Thus, the driver has the responsibility of the right use of the functions and he/she has to switch on or off the functionalities depending on the situation. However, if one wants to create now a safety relevant device, these two main characteristics will have to be drastically modified. Hence, a framework has been created to integrate safety maneuver dedicated algorithms that can take the initiative to support the driver. By speaking about spontaneity and partial solving, the concept of multi-agent appears automatically in the discussion. It has been chosen here too because it also gives the possibility to re-use standard components and to upgrade its knowledge just by adding new functionalities.

To pretend having a safety relevant framework, a redundancy mechanism has been integrated within this framework. Similarly to the airplane industry, only the full redundancy (algorithms and independent teams) can prevent a general failure. However, due to cost limitations, it is not economically possible to have redundant teams performing the same work. Therefore algorithms coming from the robotic field have been integrated parallel to the automotive controllers. Even if they do not have the same dynamics as the automotive controllers, they will have anyway a positive influence on the safety requirements. The second aspect to work on is the ability to integrate the mechanisms to switch on or off the agents. If the agents can take their initiative, they need to integrate working range patterns to determine whether their actions are compatible to the situation or not. Thus, too restrictive patterns may lead to a situation where no automotive agents switch on themselves. Here the use of robotic agents will enable the virtual driver to give advices anyway by relying on their knowledge. Actually robotic algorithms such as path planning can work every time as they are not pattern based.

The new integration concept of the assistant systems will generate the creation of a safety envelope, which includes all the safe vehicle commands. In practice, synchronized agents are used to define the optimal acceleration and steering angle for each possible maneuver. Moreover, limits of acceleration and steering are used to avoid collision with other road users. As the results from the robotic agents and from the automotive agents are different, a fusion is mandatory. First this fusion generates clusters of optimums representing the same maneuvers. After that the corresponding local safety envelopes have been fused together too. Roughly speaking, the overlapping ranges have been used as kernel for the safety envelope of a maneuver and its range can grow up depending on the confidence of the results.

The robotic agents used here are a potential field based algorithm to determine the optimal command and a dynamic window based algorithm to determine the safety limits. In parallel the automotive agents have been sorted like within a driver cognition model. Reactive and anticipating agents will compete to determine if a maneuver is currently safe and will not be dangerous in a few seconds (e.g., a passing maneuver just before a strong curve). The anticipation algorithm receives the model of the road ahead the vehicle to determine where the curvature is critical. Indeed, it can determine the current maximal allowed acceleration with still the possibility to brake enough in order to reach the maximal speed at this critical point. In parallel, the reactive part of the automotive agents will base on an environment characterization. It determines the minimal, maximal allowed speeds and distances, the lane changing possibility depending on the weather, the vehicle type and the current driving rules (e.g., time, country, type of road). After that, the reactive automotive agents can take their initiative to model the following maneuvers: distances to road users ahead or behind the ego vehicle, speed control, avoiding right passing (except during emergency cases). Hence, more complex maneuvers such as vehicle passing have been modeled by combining these basic maneuvers, for example, the lane changing synchronized with the distance control(s).

6.5 Concept of adaptive cooperation

The main idea of cooperation is the optimization of the vehicle command at the very last moment. This way of thinking is right now the single one allowed for vehicles on the road. The first functionalities introduced on the serial vehicles are only managing one command axle (acceleration/braking or lateral control). For example, as soon as the driver brakes too heavily, the command is no more understood as braking pads position but as level of deceleration, which requires a slip control. Or as soon as the minimal safe distance to the vehicle in front has been overwritten, the vehicle will engage an emergency brake to mitigate the collision.

Hence the generalization of the intervention process implies a mandatory synchronization of both acceleration/braking and lateral control. By analogy, the intervention process can only start once the limits of the safety envelope has been reached and the intervention process has to make the vehicle command reach the optimum command of the maneuver chosen by the driver.

Thus, an algorithm for maneuver detection has been integrated first within the decision control. The first required a priori knowledge is based on the choice of maneuver, the driver state and the environment to estimate the maneuver. And secondly an in situ correlation determines the probability that the driver's command tends to these maneuvers. The main problem here is the latency time allowed by the driver. A false detection or a delayed detection will generate a not-adapted feedback, which can disturb him/her.

The intervention itself is based on a transition path from the current driver's command and the optimum of his/her maneuver. This transition path has to be encapsulated within the safety envelope to ensure a safe transition.

A direct meshing algorithm has been chosen as basics for the determination of the path transition as it can be integrated within an embedded platform without risk of stack overflow (coming mostly from uncontrolled iteration processes). By using the properties of the safety envelope and the optimum included within this envelope, a direct general methodology has been found by re-using the meshing concepts from the Delaunay algorithm.

The transition will be performed after that by defining a temporary vehicle command generated by using a vehicle controller, which will be shifted step by step over the transition path. However, this method is only based on a geometrics-based sub-optimal transition and it cannot ensure in advance the stability and controllability of the transition. Hence combined with a Lyapunov function, it may provide here improved solution once it is possible to compute it in real time and within memory-protected chips.

A direct application of the transition over the safety envelope will be the integration of a pre-crash capacity. As soon as the limits of the safety envelope are overridden due to unforeseen road-users or vehicle instability, an accident cannot be avoided anymore because of collision or road departure. Hence a crash mitigation procedure can be started with the airbags' opening preparation and the seat belts tightening.

The bottleneck of this method is the hypothesis that the safety envelope will ever be determined with 100% accuracy. This implies that the virtual driver will ever see all the road-users and understand their behaviors, the vehicle state and dynamics will be 100% determined in advance (with of course the road irregularities effect) and the virtual driver knowledge will always be able to answer correctly to the situation. In practice such level of details cannot be reached. Thus, the definition of limits for the safety envelope is not as accurate as it was supposed. That is why the intervention process cannot act binary as proposed right now. In practice, the virtual driver can have a high confidence level only for standard well-defined situations.

As neither the driver nor the virtual driver can always get the whole amount of information, their levels of confidence have to be taken into account in order to adapt the cooperative way depending on who will have the most probable chance to react correctly to the situation. In practice, the driver's state is used as confidence level by looking at his/her drowsiness and his/her current activity. Here the drowsiness has been measured indirectly by looking at the command performance measurement. However, having an adaptive cooperation with active support will influence this performance. Hence only a direct method can be used for the drowsiness measure, for example, with internal camera. For the virtual driver, its confidence value results from the confidence values of the data fusion and the matching rate of the assistant systems patterns on the environment.

Once each actor (driver and virtual driver) provides the optimal vehicle command corresponding to the maneuver chosen by the driver with a corresponding confidence level, the driving task is analog to an optimization process based on a Delphi method. This concept uses several actors who provide their solutions to the given problem with their confidence value. After a combination of the solutions, a feedback is sent to each actor to let him/her have the possibility to change or to confirm his/her solution (and thus modifying his/her own confidence value). After a second round, the final solution is obtained by fusing all the provided solutions. One can see the driving task with several analogies. The two actors provide their solutions with their confidence values. As the assistant systems were supposed to have a 100% confidence value, their solutions will always be chosen once the intervention process has to start. Moreover, as their cognition was supposed to be perfect, they never have to get any feedback from the driver side. At the opposite, the driver has received a feedback from the virtual driver in order to let him/her improve his/her command.

Thus, a feedback to the virtual driver is required anyway. In practice, regions of interest have been extracted from the driver's side by looking at driver's regions of interest (looking where the driver is looking) and his/her acceptance of the feedback. These advices have provided hypothesis of missed information that are checked by different sensors (e.g., lane detection by the camera, objects by the radars sensors).

6.6 Results and next steps

The concepts presented here have been elaborated and developed during the European SPARC project. The overall architecture has been taken as described and integrated within full drive-by-wire heavy good vehicle and passenger car. The use of real-time identification has been simulated to prepare the executive platform but standard reactive stability controls have been integrated because of data inaccuracy. Hence for the rest of the verification, only the road–tire friction estimation has been used in real time to adapt the vehicle model. The simulation of the platform as first step has validated the overall concept anyway and it will be used once the sensors will be upgraded. Furthermore the road–tire estimation has shown enough stability to get potential for serial development as it proposes a low cost solution.

The framework created to integrate the assistant systems has been intensively tested in virtual reality platform and in vehicles. Thus, it can be re-used easily to integrate new functionalities and to propose a safety relevance behavior. This overhead is currently preventing the use of this technology in embedded platform. However, the standard platforms in vehicles can already integrate a static version of this framework with the automotive assistant systems and with a computation time lower than 1.3 ms. Finally the intervention process has huge possibility of improvement to make it robust to every possibility. Here a pre-verification of the intervention based on robust control may be mandatory.

7

General conclusion

Many statistics show the main role of the driver in case of accidents. Thus, research focuses on driver tasks on both executive and command levels to improve the road safety. Three parallel ways of improvement are possible. First, the road infrastructure can be improved. Secondly, devices can help the driver by monitoring his/her alertness and his/her vehicle command. Third, the executive platform can be improved to perform better the driver's command. Most researches try to solve all the problems with assistant systems. Unfortunately such systems cannot ensure a 100% safe road, because both software and hardware can have failures or even the algorithms may not be adapted to a situation.

Therefore a concept of adaptive cooperation between driver, assistant system and vehicle may be better to take the best from both driver and assistant system depending on the situation and their capacity. Such concept has been developed here with the re-design of the whole vehicle architecture to be well adapted to such technology. After analysis of the driver cognitive model, requirements to the assistant system have been defined. Like every drivers, the assistant system gets different sources of knowledge, which are used depending on the situation at the moment. The fusion of these different tasks has been the foundation for the concept of an assistant system as multi-agent system. Moreover, fail-tolerance of the assistant system can also be performed by using this multi-agent system. Indeed, redundancy of the agents can lower the risk of design failure. Here several agents have been integrated from the robotic such as the potential field approach or a modified dynamic window. From the automotive world, agents have reused the standard longitudinal controllers (e.g., speed and distance controls) and the lateral controllers (e.g., lane keeping and lane changing). Combinations of these agents behind a dedicated middle-agent have made possible a reuse of their functionality for different maneuvers. In addition, a preventive speed adaptation has been also integrated.

Nevertheless this assistant system has to feel the vehicle dynamics like the driver does, in order to define its safety envelope. An optimum can be defined for each maneuver, but the safety ranges in terms of acceleration and steering angle also needs to be defined. Therefore, the executive level sends continuously a feedback about the vehicle capacities to the command level. This feedback describes the acceleration, braking and steering capacities of the vehicle as transfer functions. Their computation is got from the estimation of μ and the position of the center of gravity on one side, and the fusion within a vehicle dynamics model of the real-time identification of the actuators on the other side.

At the end, the driver comes from one side with his/her command, and the assistant system comes from the other side with its safety envelope. If the assistant system shall help the driver, it will have to first understand what the driver wants to do. After that, the level of intervention has to be weighted by looking first at the capacity of the driver and assistant system. Actually, the capacities of both actors have to be determined by use of the driver's level of alertness and the system confidence value. After that, it is possible to determine how good it is possible to help the driver. Thus on one side, a fit driver may override the feedback from the assistant system, if it gets a bad confidence value. On the other side, the assistant system may perform completely the vehicle stabilization after having looked at the wish of a drowsy driver. Unfortunately, conflicts between driver and assistant system may happen. Most of the time, they are due to a lack of relevant information from one side. Thus, these conflicts can be solved for the driver side with an adequate feedback. However, such feedback needs also to be given to the assistant system, which also has to verify its sensing. Nevertheless the source of conflicts with the driver can only be defined here as hypotheses as many reasons for the driver can bring the same results.

With the system architecture presented here, it will be possible to integrate later improved agents dedicated to new scenarios. This integration may also require upgraded ontology to describe new concepts. Furthermore, the detection of the driver's maneuver is crucial to provide an adequate feedback to the driver. Thus, the algorithm for the detection of the maneuver transition has to be improved to lower latency time without false-positive. Finally, the cooperation with the driver stays the most complex part of the whole architecture for the future investigations. Indeed, improvements of the estimation of intervention level required from the driver have to be done and tested massively over long periods.

References

1. T. Akerstedt and S. Folkard. A quantitative model for vigilance regulations and its application in transport operations. *Vigilance and Transport Conference*, 1993.
2. L. Andreone, A. Amditis, E. Deregibus, S. Damiani, D. Morreale, and F. Bellotti. Beyond context-awareness: Driver-vehicle-environment adaptivity. from the comunicar project to the aide concept. In *IFAC 16th World Congress*, July 2005.
3. Nicholas Apostoloff and Alex Zelinsky. *Experimental Robotics VIII*, chapter Vision in and out of vehicles: integrated driver and road scene monitoring. Springer-Verlag, 2002.
4. Kai O. Arras, Jan Persson, Nicola Tomatis, and Roland Siegwart. Real-time obstacle avoidance for polygonal robots with a reduced dynamic window. 2002.
5. A. Avizienis and L. Chen. On the implementation of n version programming for software fault tolerance during program execution. In *Proceedings of the COMPSAC*, pages 149–155, 1977.
6. L. Baindridge. Ironies of automation. *Automatica*, 19:775–779, 1983.
7. Andrea Baraldi and Flavio Parmiggiani. An investigation of the textural characteristics associated with gray level co-occurrence matrix statistical parameters. In *IEEE Transactions on Geoscience and Remote Sensing*, volume 33, pages 293–304, March 1995.
8. E. Bekiaris, S. Petica, and K. Brookhuis. Driver needs and public acceptance regarding telematic in-vehicle emergency control aids. In *Proceedings of the 4th Word Congress on Intelligent Transport Systems*, pages 223–231. The Automatic Division of the Institution of Mechanical Engineers, Professional Engineering Publisher, 1997.
9. Mario Bellino, Yuri Lopez de Meneses, Peter Ryser, and Jacques Jacot. Lane detection algorithm for an onboard camera. *SPIE proceedings of the first Workshop on Photonics in the Automobile*, 2004.
10. Massimo Bertozzi and Alberto Broggi. Gold: a parallel real-time stereo vision system for generic obstacle and lane detection. In *IEEE Transactions on Image Processing*, volume 7, pages 62–81, 1998.
11. Massimo Bertozzi, Alberto Broggi, Gianni Conte, and Alessandra Fascioli. Obstacle and lane detection on the argo autonomous vehicle. *proceedings of IEEE Intelligent Transportation Systems conference*, 1997. Boston, MA.

12. Massimo Bertozzi, Alberto Broggi, Cellario Massimo, Alessandra Fascioli, Paolo Lombardi, and Marco Porta. Artifical vision in road vehicles. In *Proceedings of the IEEE*, volume 90 of 7, pages 1258–1270, 2002.
13. L. J. Blincoe and B. M. Faigin. The economic cost of motor vehicle crashes, 1990. Technical Report DOT-HS-807-876, US Department of Transportation, National Highway Traffic Safety Administration, 1992.
14. O. Bohien, S. Buller, R. W. De Doocker, and G. Aachen. Impedance based battery diagnosis for automotive applications.
15. Guy Boy. Human-center automation of transportation systems. *AAET Conference*, February 2005.
16. Guy Boy and Thomas R. Gruber. Intelligent assistant systems: Support for integrated human-machine systems. *AAAI Symposium on Knowleedge-Based Human-Computer Communication*, March 1990.
17. F. V. Brasileiro, P. D. Ezhilchelvan, S. K. Shrivastava, N.A. Speirs, and S. Tao. Implementing fail-silent nodes for distributed systems. *IEEE Transactions on Computers*, 1996.
18. Alberto Broggi, Massimo Bertozzi, Alessandra Fascioli, Corrado Guarino lo Bianco, and Aurelio Piazzi. The argo autonomous vehicle vision and control systems. In *International Journal of Intelligent Control and Systems*, volume 3, pages 409–441, 1999.
19. K. A. Brookhuis and D. De Waard. The use of psychophysiology to assess driver status. *Ergonomics*, 36:1099–1110, 1993.
20. Rodney A. Brooks. A robust layered control system, for a mobile robot. *IEEE Journal of Robotics and Automation*, RA-2(1):14–23, March 1986.
21. Robert Grover Brown and Patrick Y. C. Hwang. *Introduction to Random Signals and Applied Kalman Filtering.* John Wiley & Sons, 1997.
22. J. B. Bullock and E. J. Krawisky. Analysis of the use of digital road maps in vehicle navigation. *IEEE Position Location and Navigation Symposium*, 1994.
23. J. R. Carr and F. P. De Miranda. The semivariogram in comparison to the co-occurrence matrix for classification of image texture. *IEEE Transactions on Geoscience and Remote Sensing*, 1998.
24. H. L. Chan and D. Suanto. A new battery model for use with battery energy storage systems and electric vehicle power systems. In *IEEE Power Engineering Society*, 2000.
25. Li-Yen Chang and F. Mannering. Analysis of injury severity and vehicle occupancy in truck- and non-truck-involved accidents. In University of Washington Department of Civil Engineering, editor, *Accident Analysis and Prevention*, volume 31, pages 579–592. Accident Analysis and Prevention, vol 31 no 5, September 1999.
26. Chew, L. Paul, Nikos Chrisochoides, and Florian Sukup. Parallel constrained delaunay meshing. *AMD - Trends in Unstructured Mesh Generation*, 220:89–96, 1997.
27. COMSIS Corporation. Human factors guidelines for crash avoidance warning devices. NHTSA Contract DTNH22-91-C-07004, 1993.
28. COMSIS Corporation. In-vehicle crash avoidance warnings: Major research needs. NHTSA Contract DTNH22-91-C-07005, 1993.
29. Carlos Canudas de Wit and Panagiotis Tsiotras. Dynamic tire friction models for vehicle traction control. *American Control Conference*, June 1998.
30. Ralph Deters. Developing and deploying a multi-agent system. In CSA, editor, *4th International Conference on Autonomous Agents*, pages 175–176, 2000.

31. Ernst D. Dickmanns and A. Zapp. Autonomous high speed road vehicle guidance by computer vision. In R. Isermann, editor, *Automatic Control - World Congress, 1987: Selected Papers from the 10th Triennal World Congress of the International Federation of Automatic Control*, pages 221–266, 1987.
32. D. F. Dinges and R. C. Graeber. Crew fatigue monitoring. *Flight Safety Digest*, pages 65–75, October 1989.
33. T. A. Dingus, L.H. Hardee, and W. W. Wierwille. Development of models for on-board detection of driver impairment. *Accident Analysis and Prevention*, 19(4):271–283, 1987.
34. Joanne Bechta Dugan and Randy Van Buren. Reliability evaluation of fly-by-wire computer systems. In *Journal of Systems and Software*, volume 25, pages 109–120. Elsevier Science Inc., April 1994.
35. Inland Transport Committee Economic Commission for Europe. Convention on road traffic, November 1968.
36. Howard E. Egeth and Steven Yantis. Visual attention: Control, representation, and time course. In *Annual Revue of Psychology*, volume 48, pages 269–297, 1997.
37. Alan Fedoruk and R. Deters. Improving fault-tolerance by replicating agents. *AAMAS*, pages 737–744, 2002.
38. Volker Feil, Ulrich Gemkow, and Matthias Stämpfle. A vehicular software architecture enabling dynamic alterability of services sets. 2001.
39. L. Festinger. A theory of social comparison processes. *Human Relations*, 7:117–140, 1954.
40. Frank O. Flemisch, Catherine A. Adams, Sheila R. Conway, Ken H. Goodrich, Michael T. Palmer, and Paul C. Schutte. The h-metapher as a guideline for vehicle automation and interaction. NASA/TM 2003-212672, NASA, October 2003.
41. European Commission for Road Safety. *European Road Safety Charter*. 2001.
42. Ana L. N. Fred and Anil K. Jain. Data clustering using evidence accumulation. 2003.
43. Ana L. N. Fred and Anil K. Jain. Robust data clustering. 2003.
44. E. A. Galperin. Pareto analysis vis-à-vis balance space approach in multiobjective global optimization. *Journal of Optimization Theory and Applications*, 98(3):533–545, 1997.
45. L. Gao, S. Liu, and R. A. Dougal. Dynamic lithium-ion battery model for system simulation. In *IEEE Trans. Comp. Pack. Tech.*, 2002.
46. M. R. Genesereth and R. E. Fikes. Knowledge interchange format, version 3.0 reference manual. Technical Report Technical Report Logic-92-1, Computer Science Department, Stanford University, 1992.
47. B. S. Gibson. Visual quality and attentional capture: A challenge to the special role of abrupt onsets. *Journal of Experimental Psychology: Human Perception and Performance*, 22:1496–1504, 1996.
48. Thomas F. Golob and Amelia C. Regan. Truck-involved crashes and traffic levels on urban freeways. Technical report, University of California.
49. Richard Grace, V.E. Byrne, D.M. Bierman, J.-M. Legrand, D. Gricourt, B.K. Davis, J.J. Staszewski, and B. Carnahan. A drowsy driver detection system for heavy vehicles. In *Proceedings of the 17th Digital Avionics Systems Conference*, volume 2, pages I36/1 – I36/8, 2001.
50. Nicola Guarino. Formal ontology and information systems. *Proceedings of Formal Ontology in Information Systems*, pages 3–15, 1998.

51. F. Gustavson. Slip-based estimation of tire-road friction. Linköping University, Sweden, 1995.
52. S. Hagg. *Multi-agent systems, methodologies and applications*, chapter A sentinel approach to fault handling in multi-agent systems, pages 190–195. 1997.
53. Jin-Oh Hahn, Rajesh Rajamani, and Lee Alexander. Gps-based real-time identification of tire-road friction coefficient. In *IEEE Transactions on Control Systems Technology*, volume 10, pages 331–343, May 2002.
54. P. A. Hancock and R. Parasuraman. Human factors and safety in the design of intelligent vehicle-highway systems (ivhs). *Journal of Safety Research*, 23:181–198, 1992.
55. Uwe Handmann, Thomas Kalinke, Christos Tzomakas, Martin Werner, and Werner von Seelen. *Computer vision for driver assistance systems*, volume 3364. SPIE, July 1998.
56. J. C. Hansen and S. A. Hillyard. Endogenous brain potentials associated with selective auditory attention. *Electroencephalography and Clinical Neurophysiology*, 49:277–290, 1980.
57. N. L. Haworth. Fatige detection in trucks in normal operations. In *International Technical Conference on the Enhanced Safety of Vehicles*, 1996.
58. H. Hecht. Fault-tolerant software for real-time applications. *ACM Computing Surveys*, 8(4):391–407, 1976.
59. E. R. Hoffmann and R. G. Mortimer. Drivers' estimates of time to collision. *Accidents Analysis Prevision*, 4:511–520, August 1994.
60. Yueng-Hsiang Huang, Matthias Roetting, Jamie R. McDevitt, David Melton, and Theodore K. Courtney. In-vehicle safety feedback. *Professional Safety*, pages 20–27, January 2005.
61. K. A. Hutton. *Transportation Research Part F: Traffic Psychology and Behaviour*, volume 4, chapter Modifying Driver Behaviour with Passenger Feedback, pages 257–269. 2001.
62. A. Itoi, R., Cilvet, M. Voth, B. Dantz, P. Hyde, A. Gupta, and W. C. Dement. Can drivers avoid falling asleep at the wheel? Technical report, AAA Foundation for Traffic Safety, Februar 1993.
63. Anil K. Jain, M. N. Murty, and P. J. Flyn. Data clustering: A review. *ACM Computation Surveys*, 31(3):264–323, September 1999.
64. W. H. Janssen, M. Wierda, and A. R. A. Van der Horst. Automatisering van de rijtaak en gebruikersgedrag: een inventarisatie en research agenda (automation of driving and user behaviour). Technical Report TNO-IZF 1992 C-16, TNO-TM, 1992.
65. Nicholas R. Jennings, Katia Sycara, and Michael Wooldridge. A roadmap of agent research and development. *Autonomous Agents and Multi-agent Systems*, 1998.
66. R. Kamnik, F. Boettinger, and K. Hunt. Roll dynamics and lateral load transfer estimation in articulated heavy freight vehicles. In *I MECH E Part D Journal of Automobile Engineering*, volume 217, pages 985 – 997, November 2003.
67. Uwe Kiencke and Lars Nielsen. *Automotive Control Systems*. Springer, 2000.
68. K. H. Kim. Distributed execution of recovery blocks: An approach to uniform treatment of hardware and software faults. *Proceeding of 4th Internation Conference on Distributed Computing Systems*, pages 526–532, 1984.
69. Albert Kircher, Marcus Uddman, and Jesper Sandin. Vehicle control and drowsiness. VTI Meddelande 922A, Swedish National Road and Transport Research Institute, 2002.

70. Ronald R. Knipling and Walter W. Wierwille. Vehicle-based drowsy driver detection: Current status and future prospects. In *IVHS American Fourth Annual Meeting*, pages 2–24, April 1994.
71. P. Kohn, A. Pauly, R. Fleck, M. Pischinger, T. Richter, M. Schnabel, R. Bartz, M. Wachinger, and S. Schott. Active steering: the ride dynamic steering system of the new 5-series. *ATZ Automobiltechnische Zeitschrift*, 105(Der Neue BMW 5er Extra):96–105, August 2003.
72. Y. Koren and J. Borenstein. Potential field methods and their inherent limitations for mobile robot navigation. 1991.
73. Janice L. Krieger. Shared mindfulness in cockpit crisis situations: An exploratory analysis. *Journal of business communication*, 42, 2005.
74. Sanjeev Kumar, Philip R. Cohen, and Hector J. Levesque. The adaptive agent architecture: Achieving fault-tolerance using persistent broker teams. *Fourth international conference on autonomous agents*, 2000.
75. Tetsuya Kuno and Hiroaki Sugiura. Detection of road conditions with ccd cameras mounted on a vehicle. In *Systems and Computers in Japan*, volume 30, pages 88–99, 1999.
76. Andrew Laine and Jian Fan. Texture classification by wavelet packet signatures. Technical Report IRI-9111375, National Science Fundation, 1991.
77. D. N. Lee. A theory of visual control of braking based on information about time-to-collision.
78. Harold A. Linstone and Murray Turoff. *The Delphi Method Techniques and Applications.* 2002.
79. H. Lunenfeld and G. J. Alexander. A user's guide to positive guidance. Technical Report FHWA SA-90-017, Federal Highway Administration, 1990.
80. Min Luo, Mohamed Abdelrahman, and Jeff Frolik. Wavelet-based sensor fusion for data having different sampling rates. *American Control Conference*, June 2001.
81. S. Makeig and M. Inlow. Lapses in alertness: coherence of fluctuations in performance and eeg spectrum. *Electroencephalography and Clinical Neurophysiology*, 86:23–35, 1993.
82. Uwe Malinowski, Thomas Köhme, Hartmut Dietrich, and Matthias Schneider-Hufschmidt. A taxonomy of adaptive user interfaces. *Proceedings of the conference on people and computers VII*, pages 391–414, 1993.
83. C. Marberger. Driver warning concept. Technical Report IST-2000-28062, AWAKE consortium, 2003.
84. J. McCarthy and P. J. Hayes. *Machine Intelligence 4*, chapter Some philosophical problems from the standpoint of artificial intelligence. Edinburgh University Press, 1969.
85. Barry A. McLellan, Sandro B. Rizoli, Frederick D. Brenneman, Bernard R. Boulanger, Philip W. Sharkey, and John P. Szalai. Injury pattern and severity in lateral motor vehicle collisions: A canadian experience. *Journal of Trauma-Injury Infection and Critical Care*, 41(4):708–713, October 1996.
86. J. A. Michon and A. Smiley. *Generic intelligent driver support.* Taylor and Francis, 1993.
87. The Motor Industry Software Reliability Association MISRA. *MISRA-C:2004 - Guidelines for the use of the C language in critical systems.* oct 2004.
88. M. Mitschke. *Dynamik der Kraftfahrzeuge.* Springer-Verlag, 1972.

89. R. Molloy and R. Parasuraman. Monitoring an automated system for single failure: vigilance and task complexity effects. *Human Factors*, 38:311–322, 1996.
90. T. A. Mondor and T. E. Lacey. Faciliititative and inhibitory effects of cuing sound duration, intensity, and timbre. *Perception and Psychophysics*, 63:726–736, 2001.
91. K. Narendra and P. Gallman. An iterative method for the identification of non-linear systems using a hammerstein model. *IEEE Transactions on Automatic Control*, 11(3):546–550, July 1966.
92. Gartner Nathan, Carrol J. Messer, K. Ajay Rathi, and Rodger J. Koppa. Traffic flow theory. Technical report, 1997.
93. National Highway Traffic Safety Administration NHTSA. Traffic safety facts 2002. Technical report, U.S. Department of Transportation, 2002.
94. Ian Niles and Adam Pease. Towards a standard upper ontology. *FOIS'01*, October 2001.
95. Donald Norman. Categorization of action slips. *Psychological review*, 88:1–15, 1981.
96. Bureau of Labor Statistics. 2002 consensus of fatal occupational injuries charts. Technical report, US Department of Labor, 2003.
97. Elth Ogston, Benno Overeinder, Maarten Van Steen, and Frances Brazier. A method for decentralized clustering in large multi-agent systems. *AAMAS*, July 2003.
98. H. B. Pacejka and R. S. Sharp. Shear force development by pneumatic tyres in steady state conditions: A review of modeling aspects. *Vehicle System Dynamics*, 20(3-4):121 – 176, 1991.
99. R. Parasuraman, M. Mouloua, and R. Molloy. Effects of adaptive task allocation on monitoring of automated systems. *Human Factors*, 38:665–679, 1996.
100. Mari Partio, Bogdan Cramariuc, Moncef Gabbouj, and Ari Visa. Rock texture retrieval using gray level co-occurrence matrix. 2002.
101. Alex Pentland and Andrew Liu. Modeling and prediction of human behavior. *Neural Computation*, 11:229–242, 1999.
102. A. Polychronopoulos and A. Amditis. Definition of awake functions, scenarios and sensors to be used. system architecture and risk analysis. Technical Report IST-2000-26086, AWAKE consortium, 2004.
103. Dean Pomerleau. Ralph: Rapidly adapting lateral position handler. In *Proceedings of IEEE Symposium on Intelligent Vehicles*, pages 506–511, 1995.
104. Vasudha Ramnath, Joo-Hwee Lim, Jean-Pierre Chevallet, and Daqing Zhang. Harnessing location-context for content-based services in vehicular systems. 2005.
105. B. Randell and J. Xu. Object-oriented software fault tolerance: Framework, reuse and design diversity. *1st PDCS2 Open Workshop*, pages 165–184, 1993.
106. J. Rasmussen. Information processing and human-machine interaction - an approach to cognitive engineering. *North Holland Series in System Science and Engineering*, 1986.
107. Pedro Rodrigues, Joao Gama, and Joao Pedro Pedroso. Hierarchical time-series clustering for data streams. 2004.
108. Eric J. Rossetter and Christian J. Gerdes. A study of lateral vehicle control under a 'virtual' force framework. In *Proceedings of the 2002 AVEC Conference*, Hiroshima, Japan, 2002.

109. John Rushby. A comparison of bus architectures for safety-critical embedded systems. Technical Report NASA/CR-2003-212161, NASA, 2003.
110. A. Shaout, M. A. Jarrah, H. Al-Araji, and K. Al-Tell. A nonlinear optimal four wheels steering controller. In *Circuits and Systems, 2000, IEEE*, 2000.
111. T. B. Sheridan. Supervisory control of remote manipulators, vehicle and dynamic processes: experiments in command and displaying aiding. *Advances in Man-Machine Systems Research*, 1:44–137, 1984.
112. T. B. Sheridan. Telerobotics, automation and human supervisory control. *MIT Press*, 1992.
113. Peter Sherman. The potential of steering wheel information to detect driver drowsiness and associated lane departure. Technical Report DTRS92-G0007, Iowa State University, September 1996.
114. S. E. Shladover. Why we should develop a truly automated highway system. In *Transportation Research Record 1651*, pages 66–73. Transportation Research Board, National Research Council, Washington D.C., 1998.
115. A. Smiley and K. A. Brookhuis. Alcohol, drugs and traffic safety. *Road users and traffic safety*, 1987.
116. J.R. Smith, F. Fatehi, C. S. Woods, and J. F. Hauer. Transfer function identification in power system applications. *IEEE Transactions on Power Systems*, 1993.
117. Neville A. Stanton and P. Marsden. From fly-by-wire to drive-by-wire: safety implications of automation in vehicles. Safety Science, 1996.
118. Neville A. Stanton and Mark S. Young. A proposed psychological model of driving automation. In Taylor and Francis, editors, *Theoretical Issues in Ergonomics Science*, volume 1, pages 315 – 331. Taylor and Francis, October 2000.
119. J. A. Stem, D. Boyer, and D. Schroeder. Blink rate - a possible measure of "fatigue". *Human Factors*, 1994.
120. D. Stewart, C. J. Cudworth, and J. R. Lishman. Misperception of time-to-collision by drivers in pedestrian accidents. *Perception*, 22(10):1227–1244, 1993.
121. T. Suetomi and K. Kido. Driver behaviour under a collision warning system – a driving simulator study. Society of Automotie Engineers - SAE 970279, 1997.
122. E. D. Sussman, H. Bishop, B. Madnick, and R. Walter. Driver inattention and highway safety. pages 40–48. Transportation Research Record 1047, Transportation Research Board, 1985.
123. Jukka Takala. Global estimates of fatal occupational accidents. *Epidemiology*, 10(5):640–646, September 1999.
124. Anya Tascillo and Ronald Milller. An in-vehicle virtual driving assistant using neural networks. *Proceedings of the International Conference on Neural Networks*, 2003.
125. D. I. Tepas and M.J. Paley. Managing alertness and fatigue in advanced ivhs. In *Proceedings of the IVHS America 1992 Annual Meeting*, pages 785–789, 1992.
126. J. Theeuwes. Perceptual selectivity for color and form. *Perception and Psychophysics*, 51:599–606, 1992.
127. Kari Torkkola, Noel Massey, Bob Leivian, Chip Wood, John Summers, and Snehal Kundalkar. Classification of critical driving events. *Proceedings of the International Conference on Mahine Learning and Applications (ICMLA)*, pages 81–85, June 2003.

128. J. R. Treat, N. S. Tumbas, S. T. McDonald, D. Shinar, R. D. Hume, R. E. Mayer, R. L. Stansifer, and N. J. Catellan. Tri-level study of the causes of traffic accidents: Final report volume i: Causal factor tabulations and assessments. DOT Publication DOT HS-805 085, Institute for Research in Public Safety, Indiana University, 1979.
129. M. Vallet, S. Fakhar, D. Olivier, and D. Baez. Detection and control of the degree of vigilance of drivers. *Proceedings of the 13th International Conference on Experimental Safety Vehicles*, 1991. DOT HS 807 990, July, 1993.
130. E. V. Velsor. *Choosing 360: A Guide to Evaluating Multi-Rater Feedback Instruments for Management Development.* Center for Creative Leadership, 1997.
131. W. B. Verwey, K. A. Brookhuis, and W. H. Janssen. Safety effects of in-vehicle information systems. Technical Report TM-96-C002, TNO Human Factors Research Institute, 1996.
132. J. Wang, P. Agrawal, L. Alexander, and R. Rajamani. An experimental study with alternate measurement systems for estimation of tire road friction coefficient. *American Control Conference*, pages 4957–4962, June 2003.
133. N. J. Ward, S. H. Fairclough, and M. Humphreys. The effect of task automation in the automotive context: a field study of an autonomous intelligent cruise control system. In *International conference on experimental analysis and measurement of situation awareness*, number November, 1995.
134. E. L. Wiener and R. E. Curry. Flight deck automation: promises and problems. *Ergonomics*, 23:995–1011, 1980.
135. W. Wierwille, L. Ellsworth, S. Wreggit, R. Fairbands, and C. Kirn. Research on vehicle-based driver status/performance monitoring: Development, validation, and refinement of algorithms for detection of driver drowsiness. Technical report, National Highway Traffic Safety Administration, December 2004.
136. W. W. Wierwille, S. S. Wreggit, and M. W. Mitchell. Research on vehicle based driver status / performance monitoring. Cooperative Agreement DTNH 22-91-Y-07266, National Highway Traffic Safety Administration, Virginia Polytechnic Institute and State university, April 1992.
137. Michael J. Wooldridge and Nicholas R. Jennings. Agent theories, architectures, and languages: A survey. 1994.
138. P. I. Wouters and J. M. Bos. Traffic accident reduction by monitoring driver behavior with in-car data recorders. *Accident Analysis and Prevention*, 32:643–650, 2000.
139. Peng Xu and Ralph Deters. Fault-management in mas. 2004.
140. F. Zijlstra and G. Mulder. *Mental Worklad Under Stress. A psychological estimation*, chapter Mental Workload: Theoretical Points-of-view and an Overview of Measurement Methods, pages 21–41. Van Gorcum, 1989.
141. Joost Zuurbier and Paul Bremmer. State estimation for integrated vehicle dynamics control. 2004.
142. Robert Mayr, Regelungsstrategien für die automatische Fahrzeugführung, Längs- und Querregelung, Spurwechsel- und Überholmanöver Springer Verlag, 2001.

Index

MIX
Papier aus verantwortungsvollen Quellen
Paper from responsible sources
FSC® C105338

If you have any concerns about our products,
you can contact us on
ProductSafety@springernature.com

In case Publisher is established outside the EU,
the EU authorized representative is:
Springer Nature Customer Service Center GmbH
Europaplatz 3, 69115 Heidelberg, Germany

Printed by Libri Plureos GmbH
in Hamburg, Germany